Losing Earth

ALSO BY NATHANIEL RICH

Fiction

King Zeno
Odds Against Tomorrow
The Mayor's Tongue

Nonfiction

San Francisco Noir: The City in Film Noir from 1940 to the Present

DISCOVERY

CO_2 COULD CHANGE OUR CLIMATE AND FLOOD THE EARTH— UP TO HERE

At the Capitol in Washington, Gordon Ma
Donald shows where sea level would
in 2030 if his theories prove corre

CONTINU

Losing Earth

The Decade We Could Have Stopped Climate Change

Nathaniel Rich

PICADOR

First published 2019 by MCD
Farrar, Straus and Giroux, New York

First published in the UK 2019 by Picador
an imprint of Pan Macmillan
20 New Wharf Road, London N1 9RR
Associated companies throughout the world
www.panmacmillan.com

ISBN 978-1-5290-1582-9

Grateful acknowledgment is made for permission to reprint the following material:
Lyrics from "The Other Side," used by permission of the song's composer,
William Dorsey. "Cassandra," copyright © 1925, 1929, and renewed 1953, 1957 by
Robinson Jeffers; from *The Selected Poetry of Robinson Jeffers* by Robinson Jeffers.
Used by permission of Random House, an imprint and division of
Penguin Random House LLC. All rights reserved.

Frontispiece photograph of Gordon MacDonald from *People* magazine, October 8, 1979.
Used under license. Robert Sherbow / Time Inc. / Getty Images.
Photograph of Rafe Pomerance on pages 12–13 licensed from
AP Photo / J. Scott Applewhite.
Photograph of James Hansen on page 87 courtesy of NASA Goddard Institute
for Space Studies.
Photograph of Rafe Pomerance and Daniel Becker at Noordwijk in 1989
on pages 124–125 used by permission of Daniel Becker.

1 3 5 7 9 8 6 4 2

A CIP catalogue record for this book is available from the British Library.

Printed and bound by CPI Group (UK) Ltd, Croydon, CR0 4YY

This is for Roman

Wisdom shouts in the street,
She lifts her voice in the square;
At the head of the noisy streets she cries out;
At the entrance of the gates in the city she utters her
sayings:
"How long, O naive ones, will you love being
simple-minded?
And scoffers delight themselves in scoffing
And fools hate knowledge?
Turn to my reproof,
Behold, I will pour out my spirit on you;
I will make my words known to you.
Because I called and you refused,
I stretched out my hand and no one paid attention;
And you neglected all my counsel
And did not want my reproof;
I will also laugh at your calamity;
I will mock when your dread comes,
When your dread comes like a storm
And your calamity comes like a whirlwind,
When distress and anguish come upon you.
Then they will call on me, but I will not answer;
They will seek me diligently but they will not find me,
Because they hated knowledge."

—PROVERBS 1:20–29

Contents

Losing Earth

Introduction: The Reckoning

Nearly everything we understand about global warming was understood in 1979. It was, if anything, better understood. Today, almost nine out of ten Americans do not know that scientists agree, well beyond the threshold of consensus, that human beings have altered the global climate through the indiscriminate burning of fossil fuels. But by 1979 the main points were already settled beyond debate, and attention turned from basic principles to a refinement of the predicted consequences. Unlike string theory and genetic engineering, the "greenhouse effect"—a metaphor dating to the early twentieth century—was ancient history, described in any intro-to-biology textbook. The basic science was not especially complicated. It could be reduced to a simple axiom: the more carbon dioxide in the atmosphere, the warmer the planet. And every year, by burning coal, oil, and gas, human beings belched increasingly obscene quantities of carbon dioxide into the atmosphere.

The world has warmed more than 1 degree Celsius since the Industrial Revolution. The Paris climate agreement—the nonbinding, unenforceable, and already unheeded treaty signed on Earth Day 2016—hoped to restrict warming to 2 degrees Celsius. A recent study puts the odds of pulling this off at one in twenty. If by some miracle we succeed, we will only have to negotiate the extinction of the world's tropical reefs, a sea level rise of several meters, and the abandonment of the Persian Gulf. The climate scientist James Hansen has called a 2-degree warming "a prescription for long-term disaster." Long-term disaster is now the best-case scenario. A 3-degree warming, on the other hand, is a prescription for short-term disaster: forests sprouting in the Arctic, the abandonment of most coastal cities, mass starvation. Robert Watson, a former chairman of the United Nations Intergovernmental Panel on Climate Change, has argued that a 3-degree warming is the realistic minimum. Four degrees: Europe in permanent drought; vast areas of China, India, and Bangladesh claimed by desert; Polynesia swallowed by the sea; the Colorado River thinned to a trickle. The prospect of a 5-degree warming prompts some of the world's preeminent climate scientists, not an especially excitable type, to warn of the fall of human civilization. The proximate cause will be not the warming itself—we won't burst in flame and crumble all to ashes—but its secondary effects. The Red Cross estimates that already more refugees flee environmental crises than violent conflict. Starvation, drought, the inundation of the coasts, and the smothering expansion of deserts will force hundreds of millions of people to run for their lives. The mass migrations will stagger delicate regional truces, hastening battles over natural resources, acts of terrorism, and declarations of war. Beyond a certain point, the two great existential threats to our civilization, global warming and nuclear weapons, will loose their chains and join to rebel against their creators.

If an eventual 5- or 6-degree warming scenario seems outlandish, it is only because we assume that we'll respond in time. We'll have decades to eliminate carbon emissions, after all, before we are locked into 6 degrees. But we've already had decades—decades increasingly punctuated by climate-related disaster—and we've done nearly everything possible to make the problem worse. It no longer seems rational to assume that humanity, encountering an existential threat, will behave rationally.

There can be no understanding of our current and future predicament without an understanding of why we failed to solve this problem when we had the chance. For in the decade that ran between 1979 and 1989, we had an excellent chance. The world's major powers came within several signatures of endorsing a binding framework to reduce carbon emissions—far closer than we've come since. During that decade the obstacles we blame for our current inaction had yet to emerge. The conditions for success were so favorable that they have the quality of a fable, especially at a time when so many of the veteran members of the climate class—the scientists, policy negotiators, and activists who for decades have been fighting ignorance, apathy, and corporate bribery—openly despair about the possibility of achieving even mitigatory success. As Ken Caldeira, a leading climate scientist at the Carnegie Institution for Science in Stanford, California, recently put it, "We're increasingly shifting from a mode of predicting what's going to happen to a mode of trying to explain what happened."

So what happened? The common explanation today concerns the depredations of the fossil fuel industry, which in recent decades has committed to playing the role of villain with comic-book bravado. Between 2000 and 2016, the industry spent more than $2 billion, or ten times as much as was spent by environmental groups, to defeat climate change legislation. A robust subfield of climate literature has chronicled the machinations of industry lobbyists, the corruption of pliant scientists, and the influence campaigns that even now continue to debase the political debate, long after the largest oil and gas companies have abandoned the dumb show of denialism. But the industry's assault did not begin in force until the end of the eighties. During the preceding decade, some of the largest oil and gas companies, including Exxon and Shell, made serious efforts to understand the scope of the crisis and grapple with possible solutions.

We despair today at the politicization of the climate issue, which is a polite way of describing the Republican Party's stubborn commitment to denialism. In 2018, only 42 percent of registered Republicans knew that "most scientists believe global warming is occurring," and that percentage has fallen. Skepticism about the scientific consensus on global warming—

and with it, skepticism about the integrity of the experimental method and the pursuit of objective truth—has become a fundamental party creed. But during the 1980s, many prominent Republican members of Congress, cabinet officials, and strategists shared with Democrats the conviction that the climate problem was the rare political winner: nonpartisan and of the highest possible stakes. Among those who called for urgent, immediate, and far-reaching climate policy: Senators John Chafee, Robert Stafford, and David Durenberger; Environmental Protection Agency administrator William K. Reilly; and, during his campaign for president, George H. W. Bush. As Malcolm Forbes Baldwin, the acting chairman of Ronald Reagan's Council for Environmental Quality, told industry executives in 1981, "There can be no more important or conservative concern than the protection of the globe itself." The issue was unimpeachable, like support for the military and freedom of speech. Except the atmosphere had an even broader constituency, composed of every human being on Earth.

It was widely accepted that action would have to come immediately. At the beginning of the 1980s, scientists within the federal government predicted that conclusive evidence of warming would appear on the global temperature record by the end of the decade, at which point it would be too late to avoid disaster. The United States was, at the time, the world's dominant producer of greenhouse gases; more than 30 percent of the human population lacked access to electricity altogether. Billions of people would not need to attain the "American way of life" in order to increase global carbon emissions catastrophically; a light bulb in every other village would do it. A 1980 report prepared at the request of the White House by the National Academy of Sciences proposed that "the carbon dioxide issue should appear on the international agenda in a context that will maximize cooperation and consensus-building and minimize political manipulation, controversy and division." If the United States had endorsed the proposal broadly supported at the end of the eighties—a freezing of carbon emissions, with a reduction of 20 percent by 2005—warming could have been held to less than 1.5 degrees.

A broad international consensus had agreed on a mechanism to achieve this: a binding global treaty. The idea began to coalesce as early as February 1979, at the first World Climate Conference in Geneva, when scientists from fifty nations agreed unanimously that it was "urgently necessary" to act. Four months later, at the Group of Seven meeting in Tokyo, the leaders of the world's wealthiest nations signed a statement resolving to reduce carbon emissions. A decade later, the first major diplomatic meeting to approve a framework for a treaty was called in the Netherlands. Delegates from more than sixty nations attended. Among scientists and world leaders, the sentiment was unanimous: action had to be taken, and the United States would need to lead. It didn't.

The inaugural chapter of the climate change saga is over. In that chapter—call it Apprehension—we identified the threat and its consequences. We debated the measures required to keep the planet within the realm of human habitability: a transition from fossil fuel combustion to renewable and nuclear energy, wiser agricultural practices, reforestation, carbon taxes. We spoke, with increasing urgency and self-delusion, of the prospect of triumphing against long odds.

We did not, however, seriously consider the prospect of failure. We understood what failure would mean for coastlines, agricultural yield, mean temperatures, immigration patterns, and the world economy. But we did not allow ourselves to comprehend what failure might mean for us. How will it change the way we see ourselves, how we remember the past, how we imagine the future? How have our failures to this point changed us already? Why did we do this to ourselves? These questions will be the subject of climate change's second chapter. Call it the Reckoning.

That we came so close, as a civilization, to breaking our suicide pact with fossil fuels can be credited to the efforts of a handful of people—scientists from more than a dozen disciplines, political appointees, members of Congress, economists, philosophers, and anonymous bureaucrats. They were led by a hyperkinetic lobbyist and a guileless atmospheric physicist who, at severe personal cost, tried to warn humanity of what was coming. They risked their careers in a painful, escalating campaign to solve the problem, first in scientific reports, later through conventional avenues of political persuasion, and finally with a strategy of public shaming. Their efforts were shrewd, passionate, robust. And they failed. What follows is their story, and ours.

It is flattering to assume that, given the opportunity to begin again, we would act differently—or act at all. You would think that reasonable minds negotiating in good faith, after a thorough consideration of the science, and a candid appraisal of the social, economic, ecological, and moral ramifications of planetary asphyxiation, might agree on a course of action. You would think, in other words, that if we had a blank slate—if we could magically subtract the political toxicity and corporate agitprop—you'd think we'd be able to solve this.

Yet we did have something close to a blank slate in the spring of 1979. President Jimmy Carter, who had installed solar panels on the roof of the White House and had an approval rating of 46 percent, hosted the signing of the Israel-Egypt peace treaty. "We have won, at last, the first step of peace," he said. "A first step on a long and difficult road." The number one film in America was *The China Syndrome*; the number one song was the Bee Gees' "Tragedy." Barbara Tuchman's *A Distant Mirror*, a history of the calamities that befell medieval Europe after a major climatic change, had been near the top of the bestseller list all year. An oil well off Mexico's Gulf Coast exploded and would gush for nine months, staining beaches as far away as Galveston, Texas. In Londonderry Township, Pennsylvania, at the Three Mile Island nuclear plant, a water filter was beginning to fail. And in the Washington, D.C., headquarters of Friends of the Earth, a thirty-year-old activist, a self-styled "lobbyist for the environment," was struggling through a dense government report, when his life changed.

Part I
Shouts in the Street
1979-1982

The mad girl with the staring eyes and long
white fingers
Hooked in the stones of the wall,
The storm-wrack hair and the screeching mouth: does
it matter, Cassandra,
Whether the people believe
Your bitter fountain? Truly men hate the truth;
they'd liefer
Meet a tiger on the road.
—ROBINSON JEFFERS, "CASSANDRA," 1948

Rafe Pomerance in 1983

1.

The Whole Banana

Spring 1979

The first suggestion to Rafe Pomerance that humankind was destroying the conditions necessary for its own survival came on page 66 of the government publication EPA-600/7-78-019. It was a technical report, bound in a coal-black cover with beige lettering, about coal—one of many such reports that lay in uneven piles around Pomerance's windowless office on the first floor of the Capitol Hill town house that served as the Washington headquarters of Friends of the Earth. In the final paragraph of a chapter on environmental regulation, the coal report's authors noted that the continued use of fossil fuels might, within two or three decades, bring about "significant and damaging" changes to the global atmosphere.

Pomerance, startled, paused over the orphaned paragraph. It seemed to have come out of nowhere. He reread it. It made no sense. Pomerance was not a scientist; eleven years earlier he had graduated from Cornell with a degree in history. He had the tweedy appearance of an undernourished doctoral student emerging at dawn from the stacks, with horn-rimmed glasses and a thickish mustache that wilted disapprovingly over the corners of his mouth. His defining characteristic was his gratuitous height, six feet four inches, which seemed to embarrass him; he stooped over to accommodate his interlocutors. His active face was prone to breaking out in wide, even maniacal grins, but in composure, as when he read the coal report, it projected concern. He struggled with technical reports. He proceeded as a historian would: cautiously, scrutinizing the source material, reading between the lines. When that failed, he made phone calls, often to the authors of the reports, who tended to be surprised to hear from him. Scientists were not used to fielding questions from political lobbyists. They were not used to thinking about politics.

Pomerance had one big question about the coal report: If the burning of coal, oil, and natural gas could invite global catastrophe, why had nobody told him about it? If there was anyone in Washington—anyone in the United States of America—who should have been aware of such a danger, it was Pomerance. As deputy legislative director of Friends of the Earth, the wily, pugnacious nonprofit that David Brower helped found after resigning from the Sierra Club a decade earlier, Pomerance was one of the nation's most connected environmental activists, on intimate terms with staffers at all levels of the legislative and executive branches. That he was as easily accepted in the halls of the Dirksen Senate Office Building as at Earth Day rallies might have had something to do with the fact that he was a Morgenthau—great-grandson of Henry Sr., Woodrow Wilson's ambassador to the Ottoman Empire; great-nephew of Henry Jr., Franklin D. Roosevelt's Treasury secretary; and second cousin to Robert, district attorney for Manhattan. Or perhaps it was simply his charisma—self-effacing and rambunctious, voluble and obsessive, with a visceral talent for rousing soliloquy, he seemed to be everywhere, speaking with everyone, *in a very loud voice*, at once. His chief obsession was air. After working as an organizer for welfare rights, he spent the second half of his twenties laboring to protect and expand the Clean Air Act, the comprehensive law regulating air pollution, drafting the language of several amendments himself. That led him to the problem of acid rain, and the coal report.

He showed the unsettling paragraph to his office-mate Betsy Agle. Had she ever heard of the "greenhouse effect"? Was it really possible that human beings were overheating the planet?

Agle shrugged. She hadn't heard about it either.

That might have been the end of it had Agle not greeted Pomerance in the office a few mornings later holding a copy of a newspaper forwarded by Friends of the Earth's Denver office.

"Isn't this what you were talking about the other day?" she asked, gesturing.

There was an article about a geophysicist named Gordon MacDonald. Pomerance hadn't heard of MacDonald, but he knew all about the Jasons, the mysterious coterie of elite scientists to which MacDonald belonged. The Jasons were like one of those teams of superheroes with complementary powers who join forces in times of galactic crisis. They were convened by the U.S. intelligence apparatus to devise novel scientific solutions to the most vexing national security problems: how to detect an incoming missile; how to predict fallout from a nuclear bomb; how to develop unconventional weapons, like high-power laser beams, sonic booms, and plague-infected rats. Some of the Jasons had federal contracts or long-standing ties to U.S. intelligence; others held prominent titles at major research universities; all were united by the conviction, shared by their federal clients, that American power should be guided by the wisdom of its superior scientific minds. The Jasons met each summer in secret, and their very existence had been a loosely guarded secret until the publication of the Pentagon Papers, which exposed their plan to festoon the Ho Chi Minh Trail with motion sensors that signaled to bombers. After Vietnam War protesters set MacDonald's garage on fire, he pleaded with the Jasons to use their powers for peace instead of war.

He hoped that the Jasons could join forces to save the world. For human civilization, as he saw it, was facing an existential crisis. In "How to Wreck the Environment," an essay published in 1968, while he was a science adviser to Lyndon Johnson, MacDonald predicted a near future in which "nuclear weapons were effectively banned and the weapons of mass destruction were those of environmental catastrophe." The world's most advanced militaries, he warned, would soon be able to weaponize weather. By accelerating industrial emissions of carbon dioxide, they could alter weather patterns, forcing mass migration, starvation, drought, and economic collapse.

In the decade since, MacDonald had grown alarmed to see humankind accelerate its pursuit of this particular weapon of mass destruction, not maliciously, but unwittingly. President Carter's initiative to develop high-carbon synthetic fuels—gas and liquid fuel extracted from shale and tar sands—was a frightening blunder, the equivalent of building a new generation of thermonuclear bombs. During spring 1977 and summer 1978, the Jasons met in Boulder at the National Center for Atmospheric Research to determine what would happen once the concentration of carbon dioxide in the atmosphere doubled from pre–Industrial Revolution levels. It was an arbitrary milestone, the doubling, but a dramatic one, marking the point at which human civilization would contribute as much carbon to the atmosphere as the planet had done in the preceding 4.6 billion years. The inevitability of the doubling was not in question; a high school student could do the arithmetic. Depending on the future rate of fossil fuel consumption, the threshold would likely be breached by 2035 and no later than 2060.

The Jasons' report to the Department of Energy, *The Long-Term Impact of Atmospheric Carbon Dioxide on Climate*, was composed in an understated tone that only enhanced its nightmarish findings: global temperatures would increase by an average of 2 to 3 degrees Celsius; Dust Bowl conditions would "threaten large areas of North America, Asia and Africa"; and agricultural production and access to drinking water would plummet, triggering unprecedented levels of migration. "Perhaps the most ominous feature," however, would be the effect on the poles. Even minimal warming could "lead to rapid melting" of the West Antarctic ice sheet, which contained enough water to raise the oceans sixteen feet.

The Jasons sent the report to dozens of scientists in the United States and abroad; to industry groups like the National Coal Association and the Electric Power Research Institute; and, within the government, to the National Academy of Sciences, the Commerce Department, the Environmental Protection Agency, the National Aeronautics and Space Administration, the National Security Agency, the Pentagon, every branch of the military, the National Security Council, and the White House.

Pomerance read about this in a state of shock that, as was the pattern with him, swelled briskly into outrage. "This," he told Agle, "is the whole banana."

He had to meet Gordon MacDonald. The scientist, the article had mentioned, worked at the MITRE Corporation, a federally funded think tank that developed national defense and nuclear warfare technology. His title was senior research analyst, which was a delicate way of saying science adviser to the national intelligence community. After a single phone call, Pomerance, a Vietnam War protester and conscientious objector, drove several miles on the Beltway to a group of anonymous white office buildings that more closely resembled the headquarters of a regional banking firm than the solar plexus of the American military-industrial complex. He was shown into the office of a brawny, soft-spoken man with a wave of glossy, silverish hair over horn-rimmed frames, who possessed more than a passing resemblance to

Alex Karras—a geophysicist trapped in the body of an offensive lineman. He extended a hand like a bear's paw.

"I'm glad you're interested in this," said MacDonald, taking in the young activist.

"How could I not be?" said Pomerance. "How could anyone not be?"

MacDonald seemed miscast as a preacher of existential doom; he was too imposing of physical bearing and too decorous of manner. A bout of polio at the age of nine had left him with a chronic limp and a passion for scientific inquiry, awoken by the months of convalescence he had spent at a Dallas clinic reading medical journals about his illness. Despite his bad leg, he started at guard for the San Marcos Academy Bears and was offered a football scholarship to Rice. Harvard offered him a scholarship with no strings attached. Upon reaching campus, he swiftly earned a reputation as a prodigy. In his twenties, he advised Dwight D. Eisenhower on space exploration; at thirty-two, he became a member of the National Academy of Sciences; at forty, he was appointed to the inaugural Council on Environmental Quality, where he advised Richard Nixon on the environmental dangers of burning coal. Now approaching his fiftieth birthday, MacDonald explained that he first studied the carbon dioxide issue when he was about Pomerance's age—in 1961, when he served as an adviser to John F. Kennedy. MacDonald had followed the problem closely ever since, with increasing alarm.

He spoke for two hours. As he traced the history of humanity's understanding of the problem, explaining the fundamental science, Pomerance grew increasingly appalled.

"If I set up briefings with some people on the Hill," asked Pomerance, "will you tell them what you just told me?"

Thus began the Gordon and Rafe carbon dioxide roadshow. Pomerance arranged informal briefings with anybody he could think of in a position of power on Capitol Hill. The men settled into a routine, with MacDonald methodically explaining the science and Pomerance interjecting the exclamation points. They were surprised to find that, though most of the offices they visited had received copies of the Jasons' report, few senior officials were familiar with its findings, let alone grasped the dystopian consequences of global warming. After conversations with the EPA, *The New York Times*, the Energy Department (which, Pomerance learned, had established an Office of Carbon Dioxide Effects two years earlier, at MacDonald's urging), the National Security Council (a senior staffer, Jessica Mathews, was Pomerance's first cousin), and the White House's Council on Environmental Quality, they at last worked their way up to the president's top scientist himself, Frank Press.

Pomerance did not fully appreciate the extent of MacDonald's standing within the highest echelons of the U.S. government until they entered Press's chambers in the Old Executive Office Building, the granite fortress that stood on White House grounds, looming over the West Wing. MacDonald and Press had known each other since the Kennedy administration, when Press had figured out how to use Geiger counters to detect the Soviet Union's underground nuclear testing program. Press was familiar with the carbon dioxide issue. In July 1977, six months after Carter took office, he had written a memo to the president explaining that unchecked fossil fuel combustion might lead to a "global climatic warming" as high as 5 degrees Celsius and "large scale crop failures." "As you know," he wrote to Carter, "this is not a new issue." But Press had concluded that "the present state of knowledge" did not justify taking action in the near term. Since then, Press had overseen the development of Carter's synthetic-fuels program.

What Pomerance had expected to be yet another casual briefing assumed the character of a high-level national security meeting. Press had summoned nearly the entire senior staff of the president's Office of Science and Technology Policy, the officials consulted on every critical matter of energy and defense strategy, who did not seem especially familiar with the climate issue. Pomerance figured it was best to let MacDonald do all the talking. There was no need to emphasize to Press and his lieutenants that this was an issue of profound national significance. The solemn mood in the office told him that this was understood.

To explain what the carbon dioxide problem meant for the future, and not just the distant future, MacDonald began by turning to the distant past—to John Tyndall, an Irish physicist who was an early champion of Charles Darwin's work and died after being accidentally poisoned by his wife with sleeping pills. In 1859, Tyndall hit upon the greenhouse effect's fundamental corollary: because carbon dioxide molecules absorbed heat, variations in its atmospheric concentration could create changes in climate. This finding inspired Svante Arrhenius, a Swedish chemist and future Nobel laureate, to deduce in 1896 that the combustion of coal and petroleum for the mass production of energy could raise global temperatures. This warming would become noticeable in a few centuries, Arrhenius calculated, or sooner if consumption of fossil fuels continued to increase.

Consumption increased beyond anything the Swedish chemist could have imagined. Four decades later, a British steam engineer named Guy Stewart Callendar calculated the effect of "throwing some 9,000 tons of carbon dioxide into the air each minute." He discovered that, at the weather stations he observed, the previous five years were the hottest in recorded history. "Man," he wrote, had become "able to speed up the processes of Nature." That was in 1939.

MacDonald's voice was deliberate and authoritative, his powerful hands conveying the force of his argument. His audience listened in bowed silence. Pomerance couldn't read them. Political appointees concealed their private opinions for a living. Pomerance couldn't. He shifted in his chair, restless, glancing between the Jason and the government suits, trying to see whether they grasped the shape of the behemoth that MacDonald was describing.

MacDonald concluded his sermon with Roger Revelle, perhaps the most distinguished of the priestly caste of government scientists who, since the Manhattan Project, had advised every president on major policy; Revelle had been a close colleague of MacDonald's and Press's since they had all served together under Kennedy. Whereas Arrhenius and Callendar, in their icy Northern European hamlets, welcomed the prospect of warmer weather, Revelle recognized that human society had been organized around specific climatic conditions that, if altered, would lead to violent disruptions. MacDonald quoted from a major paper Revelle and Hans Seuss had published in 1957: "Human beings are now carrying out a large-scale geophysical experiment of a kind that could not have happened in the past nor be repeated in the future." The following year, Revelle helped the Weather Bureau establish a continuous measurement of atmospheric carbon dioxide at a site perched near the summit of Mauna Loa on the Big Island of Hawaii, 11,500 feet above the sea—a rare pristine natural laboratory on a planet blanketed by fossil fuel emissions. A young geochemist named Charles David Keeling charted the data. Keeling's graph came to be known as the Keeling curve, though it more closely resembled a jagged lightning bolt hurled toward the firmament. As MacDonald's imperturbable audience looked on, he traced the Keeling curve in the air, his thick forefinger jabbing toward the ceiling.

With each passing year, MacDonald explained, humanity's large-scale geophysical experiment grew more audacious. After Keeling had charted it for nearly a decade, Revelle shared his concerns with Lyndon Johnson, who included them in a special message to Congress two weeks after his inauguration. Johnson explained that his generation had "altered the composition of the atmosphere on a global scale" through the burning of fossil fuels. His administration commissioned a study by the president's Science Advisory Committee, led by Revelle, which warned in its 1965 report of the rapid melting of Antarctica, rising seas, increased acidity of fresh waters—changes that could be "not controllable through local or even national efforts." Nothing less than a coordinated global effort would be required. Yet no such effort materialized, and emissions continued to rise.

At this rate, said MacDonald, they could see a snowless New England, the swamping of major coastal cities, a 40 percent decline in national wheat production, and the forced migration of one-quarter of the world's population. Not within centuries—within their own lifetimes.

"And what," asked Press, "would you have us do about it?"

President Carter's efforts, in the wake of the Saudi oil crisis, to promote solar energy—he had proposed that Congress enact a "national solar strategy" and installed thirty-two solar panels on the roof of the White House to heat the First Family's water—was a strong start, said MacDonald, though Carter's plan to stimulate production of synthetic fuels was a dangerous lurch in the direction of auto-annihilation. Nuclear power, despite the recent horrors at Three Mile Island, should be expanded. But even natural gas and ethanol were preferable to coal. There was no way around it: coal production would ultimately have to end.

Carter's advisers asked respectful questions but Pomerance couldn't tell whether they were persuaded. The men stood and shook hands and Press led MacDonald and Pomerance out of his office. As they emerged onto Pennsylvania Avenue, Pomerance turned to MacDonald.

Knowing Press as you do, asked Pomerance, what do you think he's going to do?

Knowing Frank as I do, said MacDonald, I really couldn't tell you.

Pomerance grew uneasy. Since meeting MacDonald, he had fixated on the science of the carbon dioxide problem and the prospect of a political solution. But with their tour of Capitol Hill concluded, Pomerance began to question how the warming of the atmosphere might touch his own life. Lenore, his wife, was eight months pregnant. They had spent a lot of time talking about their hopes for the future. Was it ethical, he wondered, to bring a child onto a planet that before much longer could become inhospitable to humanity? Was there still time to avoid the worst? And why had it fallen to him, a thirty-two-year-old lobbyist without scientific training, to bring attention to an urgent, global crisis?

After several weeks, MacDonald called to tell him that Press had taken action. On May 22, Press wrote to the president of the National Academy of Sciences, Philip Handler, requesting a full assessment of the carbon dioxide problem. Handler tapped Jule Charney, the father of modern meteorology, to gather the nation's top oceanographers, atmospheric scientists, and climate modelers. They would judge whether MacDonald's alarm was justified—whether the world was, in fact, headed to cataclysm.

Pomerance was relieved to hear it, but he couldn't help wondering why it had taken so long. Scientists at the highest levels of government had known about the dangers of fossil fuel combustion for decades yet had produced little besides journal articles, academic symposiums, and technical reports. Nor had any politician or environmentalist championed the issue. No one had done much of anything. That, he figured, was about to change. If Charney's elite group confirmed that human civilization was hastening its own extinction, the president would be forced to act.

2.

Mirror Worlds

Spring 1979

In the living room of James and Anniek Hansen, under a bright window giving on to Morningside Park, there was a brown velvet love seat that nobody ever sat in. Erik, their two-year-old son, was forbidden to go near it. The ceiling above the couch sagged ominously, as if pregnant with some alien life form, and the bulge grew with each passing week. Jim promised Anniek that he would fix it, which was only fair, because it had been on his insistence that they gave up the prospect of a prewar apartment in Spuyten Duyvil overlooking the Hudson River and moved to this two-story walk-up with crumbling walls, police-siren lullabies, and gravid ceiling. Jim had resented the commute to the NASA Goddard Institute for Space Studies in Manhattan, complaining that such a profligate waste of his time—forty-five minutes *each way*—would soon be unsustainable, once the *Pioneer* spacecraft reached Venus and began to beam back data. But despite living within a few blocks of his office, Jim couldn't find time for the ceiling, and after four months it finally burst, releasing a confetti of browned pipes and splintered wood.

That was April. Jim repeated his vow to fix the ceiling the next time he had a spare moment. That would come, according to his calculations, on Thanksgiving Day. Anniek held him to his word, though it meant that she had to live with a hole in her living room ceiling for seven months—seven months of plaster and dust powdering the love seat.

Another promise Jim made to Anniek: he would be home for dinner every night by seven o'clock. By half past eight, however, he was back to his mathematical preoccupations. Anniek did not begrudge him his deep commitment to his research; it was one of the things she loved about him. Still it baffled her that the subject of his obsession should be the atmospheric conditions of a planet more than twenty-four million miles away. It baffled Jim, too, when he came to think about it. How he had traveled to Venus from Denison, Iowa, where he had been the youngest child of a diner waitress and an itinerant farmer turned bartender, was a mystery, the outcome of a series of bizarre twists of fate for which he claimed no agency. It was just something that had happened to him.

Hansen figured he was the only NASA scientist who, as a child, did not dream of outer space. He dreamed only of baseball. On clear nights, his transistor radio picked up the broadcast of the Kansas City Blues, the New York Yankees' Triple-A affiliate. Every morning, he cut the box scores out of the *Omaha World-Herald* (for which he served as chief Denison delivery boy from third grade through high school), pasted them into a notebook, and tallied statistics. In a childhood of deprivation and meagerness—during his earliest years he shared two rooms with six siblings in homes that lacked running water and a refrigerator—Hansen found comfort in numbers. He majored in math and physics at the University of Iowa. But he never would have taken an interest in celestial matters were it not for the unlikely coincidence of two events during his graduation year: the eruption of a volcano in Bali and a total eclipse of the moon.

On the penultimate night of 1963—whipping wind, 12 below—Hansen accompanied his astronomy professor to a cornfield miles outside of town. They set a telescope in an old corncrib that, Hansen shortly discovered, was being used as shelter by every beetle, fly, and wasp from the surrounding forty acres. Between two and eight o'clock in the morning, Hansen made continuous photoelectric recordings of the eclipse, pausing only when the extension cord froze and when he had to race to the car to avoid frostbite.

During an eclipse, the moon resembles a tangerine or, if the eclipse is total, a drop of blood. But this night, to the consternation of Hansen's professor, the moon vanished altogether. Hansen made the mystery the subject of his master's thesis, concluding that the moon had been obscured by the dust erupted into the atmosphere by Mount Agung, on the other side of the planet from his corncrib, six months earlier. The discovery stirred in him a fascination with the influence of invisible particles on the visible world. You could not make sense of the visible world, he realized, until you understood the whimsies of the invisible one.

A leading authority on the invisible world happened to be teaching at the University of Iowa: James Van Allen, who had grown up in a nearby farming town, had designed the first U.S. satellite, chaired the team of scientists that proposed sending people to the moon, and made the first major discovery of the space age, identifying the two doughnut-shaped regions of convulsing particles that circle Earth, later known as the Van Allen belts. At Van Allen's prodding, Hansen turned from the moon to Venus. Why, he tried to determine, was its surface so hot? In 1967, a Soviet satellite beamed back the answer: the planet's atmosphere was mainly carbon dioxide. Though once it might have had habitable temperatures, it appeared to have succumbed to a runaway greenhouse effect. As the sun grew brighter, Venus's ocean, believed to have covered the planet by an average depth of eighty feet, began to evaporate, thickening the atmosphere, which forced yet greater evaporation—a self-perpetuating cycle that finally boiled off the ocean entirely and heated the planet's surface to more than 800 degrees Fahrenheit. At the other extreme, Mars's threadbare atmosphere had insufficient carbon dioxide to trap much heat at all, leaving it about 900 degrees colder. Earth lay in the middle, its Goldilocks greenhouse effect just strong enough to support life.

Anniek expected Jim's professional life to resume some semblance of normality once the data from Venus had been collected and analyzed. But shortly after *Pioneer* entered

Venus's atmosphere, Hansen came home from the office in a state of uncharacteristic fervor, with exciting news—and an apology. The prospect of two or three more years of intense work had sprung open before him. NASA was expanding its study of Earth's atmospheric conditions. Hansen had already done some research on the global atmosphere while developing weather models for Jule Charney, who had visited the Goddard Institute. Now Hansen would have the opportunity to apply to Earth the lessons he had learned from Venus.

We want to learn more about our climate, he told Anniek—and how human beings can influence it. They would use giant new supercomputers to map the planet's atmosphere. With software programs they would create Mirror Worlds: parallel realities that mimicked our own. Everything that happened on Earth was subject to the laws of physics, represented by mathematical formulas. Most of these formulas had been developed decades, if not centuries, earlier. But it was not until the refinement of supercomputers in the 1950s and 1960s that the formulas governing the behavior of the sea, land, and sky could be combined into a single computer model. The Mirror Worlds, technically "general circulation models," could predict such complex phenomena as regional weather patterns, storm formation, vegetation growth, and the dynamics of ocean circulation. The Mirror Worlds differed significantly from the real world in just one regard: they could be sped forward to reveal the future.

Anniek's disappointment—another several years of distraction, stress, time spent apart from family—was tempered, if only slightly, by the high strain of Jim's enthusiasm. It wasn't often that her husband seemed giddy. Busy, sure;

passionate, yes. But this was different. She thought she understood it.

So this means, she asked, that you might figure out a way to predict weather more accurately?

Yes, said Jim. Something like that.

3.

Between Clambake and Chaos

July 1979

The scientists summoned by Jule Charney to judge the fate of civilization arrived on July 23, 1979, with their wives, children, and weekend bags at a three-story mansion in Woods Hole, on the southwestern spur of Cape Cod. They would review the available science and decide whether the White House should take seriously Gordon MacDonald's prediction of a climate apocalypse. The Jasons had predicted a warming of 2 or 3 degrees Celsius by the middle of the twenty-first century, but like Roger Revelle before them, they emphasized reasons for uncertainty. Jule Charney—warmhearted, dashing, gregarious—asked his scientists to quantify that uncertainty. They had to get it right:

their conclusion would be delivered to the president. But first, Charney announced, they would hold a clambake.

They gathered with their families on a bluff overlooking Quissett Harbor and took turns tossing mesh produce bags stuffed with lobster, clams, and corn into a bubbling cauldron. They exchanged pleasantries and admired the sunset, the water twinkling between the masts of moored Herreshoffs. It had been an unseasonably hot day, in the high 80s, but the harbor breeze was salty and cool. It didn't look like the dawning of an apocalypse. It looked more like a family reunion. While the children scrambled across the rolling green lawn, the scientists mingled with a claque of visiting dignitaries, whose status lay somewhere between chaperone and client—men from the Departments of State, Energy, Defense, and Agriculture; the EPA; and the National Oceanic and Atmospheric Administration. The government officials, many of them scientists themselves, tried to suppress their awe of the legends in their presence: Henry Stommel, the world's leading oceanographer; his protégé, Carl Wunsch, a Jason; the Manhattan Project alumnus Cecil Leith; the Harvard planetary physicist Richard Goody. These were the men who, in the last three decades, had discovered foundational principles underlying the relationships between the sun, atmosphere, land, and ocean—which was to say, the climate.

The hierarchy was made visible during the workshop sessions, held in the carriage house next door: the scientists sat at tables arranged in a rectangle, while their federal observers sat along the room's perimeter, observing the action as if at a theater in the round. The first two days of meetings didn't make good theater, however, as the scientists reviewed the basic principles of the carbon cycle, ocean circulation, and radiative transfer. On the third day, Charney introduced a new prop: a black speaker attached to a telephone. He dialed, and Jim Hansen answered.

Charney called Hansen because he had grasped that in order to determine the exact range of future warming, his group would have to venture into the realm of the Mirror Worlds. Charney himself had used a general circulation model to revolutionize weather prediction. But Hansen was one of just a few climate modelers who had studied the effects of carbon emissions. When, at Charney's request, Hansen programmed his model to evaluate a future of doubled carbon dioxide, it projected a temperature increase of 4 degrees Celsius. That was twice as much warming as the prediction made by the most prominent climate modeler, Syukuro Manabe, whose government lab at Princeton was the first to model the greenhouse effect. The difference between the two predictions—between a warming of 2 degrees Celsius and 4 degrees Celsius—was the difference between damaged coral reefs and no reefs whatsoever, between thinning forests and forests choked by desert, between catastrophe and chaos.

In the carriage house, the quiet, disembodied voice of Jim Hansen explained how his model weighed the influences of clouds, oceans, and snow on warming. The elder scientists interrupted, shouting questions; when they did not transmit through the telephone, Charney repeated them in a bellow. The questions kept coming, often before their younger respondent could finish his answers, and Hansen wondered if it wouldn't have been faster for him to drive the five hours and meet with them in person.

In the end Charney left it to Akio Arakawa, a pioneer of computer modeling and the world's leading authority on clouds, to determine which prediction was more accurate. On the final night at Woods Hole, Arakawa stayed up late in his motel room, printouts from Hansen and Manabe blanketing his double bed. The discrepancy, Arakawa concluded, came down to ice and snow. The whiteness of the world's snowfields reflected sunlight; if a warmer climate caused more ice to melt, less radiation would escape the atmosphere, leading to even greater warming. Shortly before dawn, Arakawa concluded that Manabe had underestimated the influence of melting sea ice, while Hansen had overemphasized it. The best estimate lay exactly in between. Which meant that the Jasons' calculation was too optimistic. When carbon dioxide doubled in 2035 or thereabouts, global temperatures would increase between 1.5 and 4.5 degrees Celsius, with the most likely outcome falling in the middle: a warming of 3 degrees.

The publication several months later of Jule Charney's report, *Carbon Dioxide and Climate: A Scientific Assessment*, was not accompanied by a banquet, a parade, or even a press conference. Yet within the highest levels of the federal government, the scientific community, and the oil and gas industry—within the commonwealth of people who had begun to concern themselves with the future habitability of the planet—the Charney report almost immediately assumed the authority of settled fact. It was the summation of all the predictions that had come before and it would withstand the scrutiny of the decades that followed. Charney's group had considered everything known about ocean, sun, sea, air, and fossil fuels and had distilled it to a single number: three. When the doubling threshold was broached, as appeared inevitable, the world would warm by 3 degrees Celsius. The last time the world was 3 degrees warmer was during the Pliocene, three million years ago, when beech trees grew in Antarctica, the seas were eighty feet higher, and wild horses galloped across the Canadian coast of the Arctic Ocean.

Still the Charney report left Jim Hansen with more urgent questions. A warming of 3 degrees would be nightmarish, but unless carbon emissions ceased suddenly, 3 degrees would be only the beginning. The real question was whether the warming trend could be reversed. Was there time to act? The report warned that "a wait-and-see policy may mean waiting until it is too late." But how would a global commitment to cease burning fossil fuels come about, exactly? Who had the power to make such a thing happen? Hansen didn't know how to begin to answer these questions. He didn't know anything about politics, after all. But he'd learn.

4.

Enter Cassandra, Raving

1979-1980

As James Hansen was charting his Mirror Worlds and Rafe Pomerance was working his connections on Capitol Hill, a small group of philosophers, economists, and social scientists were busy conducting a vigorous debate, largely among themselves, about whether a human solution to this human problem was even possible.

These scholars—call them the Fatalists—did not trouble themselves about the details of warming; they took the worst-case scenario as a given. Nor did they concern themselves with whether humanity could cease burning fossil fuels within some fixed period of time; they assumed that a solution was technically possible. They asked instead whether human beings, when presented with this particular existential crisis, were *willing* to prevent it.

It was not so simple a question as it appeared. In the middle of the eighteenth century, when fossils were first burned to generate energy on an industrial scale, an unprecedented disjunction occurred in the course of civilization. Humanity lost control of its technology. The new, world-moving inventions—the spinning jenny, the coke-fueled furnace, the coal-fed steam engine—invited dangers that their creators had not anticipated and, increasingly, could not avoid. The black smoke erasing daylight from London and Yorkshire offered an early example of unintended consequences; the Dust Bowl revealed that the short-term benefits of mechanization could lead to the frivolous discounting of ancient wisdom; and the wide adoption of gasoline-powered automobiles showed the power of technological advancement to breed mass delusion, as in 1943, when residents of Los Angeles, swimming in smog, believed the city to be under chemical attack from the Japanese. The perils increased in proportion to the power of the technology until, by the nuclear age, it became possible for the species to commit suicide as easily as pressing a button. In a 1977 report prepared by the National Research Council, Roger Revelle and Charles Keeling argued that carbon emissions posed an equal threat. "It has become increasingly apparent in recent years," they wrote, "that human capacity to perturb inadvertently the global environment has outstripped our ability to anticipate the nature and extent of the impact."

The critical word was *inadvertently*. Effect had been severed from cause. As our technology grew more sophisticated, our behavior grew more childish. Though many of our routine activities required the combustion of vast quantities of carbon, we were only passively aware, at the periphery of our consciousness, of the hum of air-conditioning, the click of a light switch, the rumble of an internal combustion motor. The debt accrued nonetheless. And the bill would come. Another report in 1977, commissioned by the Energy Research and Development Administration (the predecessor to the Energy Department), warned that humanity's fossil fuel habit would lead inexorably to a host of "intolerable" and "irreversible" disasters, but it categorized the best available remedy—a transition to renewable energy—as far-fetched. "Any government action requires political consensus," concluded the authors. "Such consensus may be difficult to achieve."

The anthropologist Margaret Mead, who knew something about the rigidity of cultural patterns, had understood the urgency of the problem even earlier, in 1975, when she convened a global warming symposium at the National Institute of Environmental Health Sciences. "We are facing a period when society must make decisions on a planetary scale," she wrote. Her conclusions were stark, immediate, and unadorned with the caveats that dominated the academic literature. "Unless the peoples of the world can begin to understand the immense and long-term consequences of what appear to be small immediate choices," she said, "the whole planet may become endangered."

But the Fatalists wondered whether greater awareness of the problem really would provoke a sensible response. Was the threat of distant catastrophe sufficient to motivate change? If so, how much threat, and how much change? We worry about our children's futures and our grandchildren's futures. But how much, precisely? And how much do we worry about our great-grandchildren or *their* great-grandchildren? Enough to compromise our living standards? An abrupt transition to renewable forms of energy called for sacrifices. Was the prospect of, say, a global food shortage one century hence enough to motivate a person to commute to work by public bus? Was it enough to convince a family of four to forgo a dryer for a clothing rack? And what degree of certainty was required if so? Thirty percent? Ninety-eight percent? The question had to be asked not only of individuals but also of nations and corporations. How much value did we assign to the future?

The answer, any economist would cede, was exceedingly little. Economics, the science of assigning value to human behavior, priced the future at a deep discount. The benefit of a short-term gain dwarfed the cost of a long-term risk. As Lester Lave, an economist at the Brookings Institution who began studying climate change in the seventies, put it at the time, "If the world were to disappear twenty-five or thirty years from now, it would make no difference to economists today." This made the threat of climate change the perfect economic disaster. By the end of the seventies, the Yale economist William Nordhaus, a member of President Carter's Council of Economic Advisers, had become so alarmed by the problem that he developed a new economic model to deal with it.

Since climate had been consistent for centuries, Nordhaus noted, human beings had taken it for granted and failed to assign it value. But a stable climate had a gargantuan monetary value. As Roger Revelle had observed, trillions of dollars' worth of long-term investments—in infrastructure, agriculture, national security, and urban development—relied on the assumption that the basic conditions governing the natural world were permanent. Jesse Ausubel, then a young staffer at the National Academy of Sciences (and one of the first people in the world to serve as a full-time climate change analyst), posed the challenge this way: "What do you do when the past is no longer a guide to the future?" Even a slightly warmer climate would incur extraordinary costs. Some scientists had already begun to try to quantify these in dollar amounts. At the National Center for Atmospheric Research, Stephen Schneider and Robert Chen, who had assisted the Charney group, had found that about five meters of sea level rise would imperil 6 percent of the nation's real estate wealth. That correlation had a limit: once the water rose beyond a certain threshold, the national economy, like the properties themselves, crashed into the sea. But the rise of the oceans would be just the beginning of the economic pain. It would be followed by agricultural decline, increased conflict between northern and southern states, an amplification of economic inequality, the dissolution of national boundaries. Nordhaus argued that it was irrational—not just morally, which was materially irrelevant anyway, but economically—to delay action.

If human behavior couldn't be improved, perhaps the market could. His remedy was to make nations pay the true cost of carbon by levying a tax on emissions. By his calculation, the price came out to ten dollars a ton. A global carbon tax, however, required a global tax collector. And that would require an international treaty.

Which raised the question: Was a strong treaty possible, even under the most favorable of circumstances? Nordhaus didn't think so. Nor did Michael Glantz, a political scientist at the National Center for Atmospheric Research. Writing in *Nature* in 1979 ("A Political View of CO_2"), Glantz observed that politicians tended to take one of two approaches to major problems: "crisis management" or "muddling through." Decisive action was only taken during a crisis, like the toxic smog that in the 1960s caused people to drop dead in the streets of major American cities and led to the passage of air pollution laws. Belated reforms were more expensive and far less effective than preventative action would have been, but that was how we addressed social problems: tardily, with half measures. Incremental dangers like air pollution were inevitably muddled through, since a society's short-term needs (unfettered energy production) eclipsed long-term environmental consequences, no matter how cataclysmic. The only future worthy of consideration was the short term. The longest term of any elected office in America was six years.

Even if the world's powers consented to negotiate a treaty, it was bound to be toothless, argued the German physicist-philosopher Klaus Meyer-Abich. Since every nation had its own set of interests, a global compromise would inevitably favor the minimal action. That was the lowest-common-denominator law of international diplomacy, and it was inflexible. But Meyer-Abich's fatalism went deeper still. The oil crisis had already proved the dangers—environmental, geopolitical, and economic—of fossil fuel combustion. If that crisis, with its dramatic, sudden negative consequences, could not convince the world to transform its energy model, the more abstract and gradual threat of climate change had no chance. Some mitigation was possible, sure, at the margins, but only measures with other short-term economic benefits. By this logic, a binding treaty to reduce emissions seemed fantastical. Put another way, the only viable approach to the dawning existential crisis of climate change was to do nothing.

The Fatalists attended the major summits, like the World Climate Conference and a major carbon dioxide symposium held by the Energy Department in spring 1979. But when they spoke, no one seemed to listen. The physical scientists who dominated such meetings simply nodded along, waiting for the opportunity to resume debating the relative influences of radiative transfer and albedo. That was their field of expertise, after all—clouds, oceans, forests, the invisible world. And so it happened that the economists, philosophers, and political scientists came to feel that, no matter how forcefully they issued their warnings, they were becoming invisible themselves.

5.

A Very Aggressive Defensive Program

1979–1980

After the publication of the Charney report, Exxon decided to create its own carbon dioxide research program, with an annual budget of $600,000. But it wanted to ask a slightly different question than Jule Charney had. Exxon didn't concern itself primarily with how much the world would warm. It wanted to know how much of the warming could be blamed on Exxon.

A manager in its research laboratory named Henry Shaw was convinced that the company needed a deeper understanding of the issue in order to influence future federal efforts to restrict emissions. "It behooves us to start a very aggressive defensive program," Shaw wrote in a memo to a supervisor, laying out his case, "because there is a good probability that legislation affecting our business will be passed."

Exxon had been tracking the carbon dioxide problem since before it was Exxon. In 1957, scientists from its predecessor, Humble Oil, published a study analyzing "the enormous quantity of carbon dioxide" contributed to the atmosphere since the Industrial Revolution "from the combustion of fossil fuels." Even then, the notion that burning fossil fuels had increased the concentration of carbon in the atmosphere went unquestioned by Humble's scientists. What was new, in the late fifties, was the effort to quantify what percentage of emissions had been contributed by the oil and gas industry. The American Petroleum Institute, the industry's largest trade association, had already begun similar studies—in 1955 it financed research by geochemists at the California Institute of Technology, who had found that fossil fuel combustion had increased the concentration of atmospheric carbon by about 5 percent.

The warnings continued. In December 1957, Edward Teller, who had led the development of the hydrogen bomb, told members of the American Chemical Society, which included engineers from oil and gas companies, that the exploitation of fossil fuels might bring about climate change; he repeated the message in 1959 at a centennial celebration of the American oil industry in New York City, organized by API and Columbia Business School. "When the temperature does rise by a few degrees over the whole globe," he told the assembled dignitaries, "there is a possibility that the icecaps will start melting and the level of the oceans will begin to rise." In 1968, an API study conducted by the Stanford Research Institute concluded that the burning of fossil fuels would bring "significant temperature changes" by the year 2000. It was "ironic," the study's authors noted, that politicians, regulators, and environmentalists fixated on incidents of air pollution that were localized and immediately observable, while the climate crisis, which would cause damage of far more daunting severity and scale, went unheeded.

The ritual repeated itself every few years. Industry scientists, at the behest of their corporate bosses, reviewed the problem, finding good reasons for alarm and better excuses to do nothing. Why should they, when almost nobody within the United States government—nor, for that matter, within the environmental movement—seemed worried? As the National Petroleum Council put it in a 1972 report prepared for the Department of the Interior, climate changes would probably not be apparent "until at least the turn of the century." The industry had enough emergencies already: antitrust legislation introduced by Senator Ted Kennedy; concerns about the health risks of gasoline; battles over the Clean Air Act; and the financial shock of benzene regulation, which increased the cost of every gallon of gas sold in America. Why take on an intractable problem that would not be detected until the current generation of employees was safely retired? Besides, the remedies seemed more punitive than the problem itself. Energy use, historically, had correlated to economic growth—the more fossil fuels we burned, the better our lives became. Why mess with that?

But the Charney report had changed the industry's cost-benefit calculus. A formal consensus about the nature of the crisis had cohered. As Henry Shaw emphasized in his conversations with Exxon's executives, the cost of inattention would rise in step with the Keeling curve.

To begin his very aggressive, defensive carbon dioxide program, Shaw turned to Wallace Broecker, a Columbia University oceanographer who was the second author of Roger Revelle's 1965 carbon dioxide report for Lyndon Johnson. In 1977, in a presentation at the American Geophysical Union, Broecker predicted that fossil fuels would have to be restricted, either by taxation or by fiat. More recently, he had testified before Congress that carbon dioxide was "the No. 1 long-term environmental problem." If presidents and senators trusted Broecker to tell them the bad news, Shaw figured that he would do for Exxon.

Broecker did not think much of Shaw's first proposal for Exxon's new program: testing the carbon levels in vintage bottles of French wine to chart the rise of atmospheric carbon dioxide over time. Broecker did agree to collaborate with a colleague, Taro Takahashi, on a more ambitious experiment conducted on board one of Exxon's largest supertankers, the *Esso Atlantic*, to determine how much carbon the oceans could absorb before coughing it back into the atmosphere. But the data came back a mess and the project was abandoned.

Shaw was running out of time. In 1978, an Exxon colleague circulated an internal memo warning that humanity had only five to ten years before "hard decisions regarding changes in energy strategies might become critical." But Congress, as Shaw had anticipated, seemed ready to act a lot sooner than that. On April 3, 1980, Senator Paul Tsongas, a Massachusetts Democrat, held the first congressional hearing on carbon dioxide buildup in the atmosphere. Gordon MacDonald testified that the United States should "take the initiative" and develop, through the United Nations, a way to coordinate every nation's energy policies to address the problem. That June, President Carter signed the Energy Security Act of 1980, which directed the National Academy of Sciences to start a multiyear, comprehensive study, to be detailed in a report called *Changing Climate*, that would analyze the social and economic consequences of climate change. Most urgently, the National Commission on Air Quality, at the request of Congress, invited two dozen experts, including Henry Shaw himself, to a meeting in Florida to develop climate legislation.

It appeared that some federal decree to restrict carbon emissions was inevitable. The Charney report had confirmed the diagnosis of the problem—a problem that Exxon helped create. Now Exxon would help shape the solution. Henry Shaw flew to Florida.

6.

Tiger on the Road

October 1980

Two days before Halloween, Rafe Pomerance traveled to a cotton candy castle in the Gulf of Mexico on a narrow spit of porous limestone that rose no higher than five feet above the sea. The Pink Palace, as locals called the Hotel Don CeSar, was a child's daydream of birthday cake dimensions, its cantilevered planes of bubble gum stucco mounted by turrets with white cupolas like melting scoops of vanilla. It stood three miles off the Suncoast amid blooms of poisonwood and gumbo limbo, and at high tide the waves came within two hundred feet of the Buena Vista Bar, where the bartender served Pink Ladies. In its carnival of historical amnesia and childlike faith in the power of fantasy, the Pink Palace was a fine setting for the first rehearsal of a conversation that would be earnestly restaged, with little variation and increasing desperation, for the next four decades.

In the year and a half since he had read the coal report, Pomerance had attended countless conferences and briefings about the science of global warming. But nobody until now had shown much interest in the only subject that he cared about, the only subject that mattered: how to *prevent* warming. In one sense he had himself to thank. During the expansion of the Clean Air Act, he had pushed for the creation of the National Commission on Air Quality, charged with ensuring that the goals of the act were met. One such goal was a stable global climate. The Charney report made clear that goal was not being met and now the commission wanted to hear proposals for legislation. It was a profound responsibility and the two dozen experts at the Pink Palace—policy gurus, deep thinkers, an industry scientist, and an environmental activist—had only three days to achieve it, but the utopian setting made everything seem possible. The conference room, with its tall windows framing postcard views of the beach, looked better suited to hosting a debutante ball than a political roundtable. The sands were a confectionary shade of white, the surf was idle, the air unseasonably hot, and the dress code relaxed: sunglasses and guayaberas, jackets frowned upon.

"I have a very vested interest in this," said State Representative Tom McPherson, a Florida Democrat, introducing himself to the delegation, "because I own substantial holdings fifteen miles inland of the coast, and any beachfront property appreciates in value."

There was no formal agenda, just a young moderator from the EPA named Thomas Jorling and a few handouts on each seat, among them the Charney report. Jorling acknowledged the vagueness of their mission.

"We are flying blind, with little or no idea where the mountains are," he said. But the stakes couldn't be higher: a failure to recommend policy would be the same as endorsing the present policy—which was no policy. "Would anyone like to break the ice?" he asked, failing to grasp the pun.

"We might start out with an emotional question," proposed Thomas Waltz, an economist at the National Climate Program. "The question is fundamental to being a human being: Do we care?"

This provoked huffy consternation.

"In caring or not caring," said John Laurmann, a Stanford engineer who had briefed Henry Shaw and other oil and gas industry scientists on the climate problem, "I would think the main thing is the timing." It was not an emotional question, in other words, but an economic one: How much did we value the future?

We have less time than we realize, said an MIT nuclear engineer named David Rose, who studied how civilizations responded to large technological crises. "People leave their problems until the eleventh hour, the fifty-ninth minute," he said. "And then: '*Eloi, Eloi, lama sabachthani?*'"

It was a promising beginning, Pomerance thought. Urgent, detailed, clear-eyed. Why had it taken so long to put a group like this together?

John Perry, a meteorologist who had worked as a staffer on the Charney report, proposed an old schoolboy trick:

they should solve the problem backward. "Suppose by some mechanism the world has managed to control the growth in atmospheric CO_2 levels by sometime in the first part of the next century," he said. "Then ask: How could that have come about?"

There was general agreement that some kind of international treaty would be needed to keep atmospheric carbon dioxide at a safe level. But nobody could agree on what that level was. And any policy that restricted the use of energy would cause trouble.

"Changes in how we are able to burn things, how we get our fuel," agreed the EPA's John Hoffman, "have tremendous destabilizing effects on society."

And if the United States did act, what good would it do? William Elliott, a NOAA scientist, introduced some hard facts: If the whole country stopped burning carbon that year, it would delay the arrival of the doubling threshold by only five years. If the entire Western world somehow managed to stabilize emissions, it would forestall the inevitable by eight years. The only way to avoid the worst was to stop burning coal. Yet China, the Soviet Union, and the United States, by far the world's three largest coal producers, were frantically accelerating extraction.

"Do we have a problem?" asked Anthony Scoville, a Republican appointee on the House science committee. "We do, but it is not the atmospheric problem. It is the political problem of the inertia of the economic and political system and the time it takes to get decisions put into effect." He doubted that it was possible for a scientist to produce a study that would convince politicians to act. The whole model was screwy. When had science alone ever forced the passage of a bill?

Pomerance found himself staring out at the beach, where the occasional tourist dawdled in the surf. Outside of the lurid pink hotel, few Americans realized that the planet would soon cease to resemble itself.

What if the problem, continued Scoville, was that they were thinking of it as a problem? Might it not be better to think about it as a *solution*—to economic stagnation, the traumas of Saudi oil, and air and water pollution? Even if the coal and oil industries collapsed, new energy technologies, like solar, would thrive, and the broader economy would be healthier for it. Now Carter was planning to invest $80 billion in synthetic-fuel development. “My God,” said Scoville, “with $80 billion, you could have a photovoltaics industry going that would obviate the need for synfuels forever!”

The talk of ending oil production stirred for the first time the gentleman from Exxon. “I think there is a transition period,” said Henry Shaw. “We are not going to stop burning fossil fuels and start looking toward solar or nuclear fusion and so on. We are going to have a very orderly transition from fossil fuels to renewable energy sources.”

“We are talking about some major fights in this country,” said Waltz, the economist. “We had better be thinking this thing through.”

But first—lunch. It was a bright day, in the low 80s, and the group voted to break for more than three hours to take advantage of the high Florida sun. Pomerance couldn't—he was restless. He had refrained from speaking, happy to let others lead the discussion, provided it moved in the right direction. But the high-minded talk had stalled into fecklessness and pusillanimity. He reflected that, as with most meetings on the subject, he was just about the only participant without an advanced degree. But few of these policy geniuses seemed to have much sense. They understood what was at stake, but they hadn't taken it to heart. They remained cool, detached—pragmatists overmatched by a problem that had no pragmatic resolution. "Prudence," Jorling had said, "is essential." But prudence was suicidal.

After lunch, Jorling tried to focus the conversation. What did they need to know in order to act?

David Slade, who as the director of the Energy Department's $200 million Office of Carbon Dioxide Effects had probably considered the question as deeply as anyone in the room, said he figured that at some point, probably within their lifetimes, they would see the warming themselves.

"And at that time," bellowed Pomerance, "it will be too late to do anything about it."

Yet nobody could agree what to do now. There was talk of disincentives to fossil fuel combustion and incentives to develop renewable energy. Anthony Scoville worried about the politicization of the issue once legislation or treaties were proposed. John Perry suggested they might request that American energy policy "take into account" the risks of global warming, though he acknowledged that a nonbinding measure might seem "intolerably stodgy."

"It is so weak," said Pomerance, the air seeping out of him, "as to not get us anywhere." His reticence was gone, replaced by a ripening frustration that threatened the fragile professional decorum that had held to this point.

Reading the anxiety in the room, Jorling reversed himself and wondered if it might be prudent to avoid proposing any specific policy. "Let's not load ourselves down with that burden," he said. "We'll let others worry about that."

Pomerance begged Jorling to reconsider. The commission had asked for hard proposals. But why stop there? Why not propose a new national energy plan? "There is no single action that is going to solve the problem," said Pomerance. "You can't keep saying, '*That* isn't going to do it,' and '*This* isn't going to do it,' because then we end up doing *nothing*."

Scoville pointed out that the United States was responsible for the largest share of carbon emissions. But not for long. "If we're going to exercise leadership," he said, "the opportunity is now." One way to lead, he proposed, would be to classify carbon dioxide as a pollutant under the Clean Air Act and regulate it as such. This was received by the room like a belch. By Scoville's logic, every sigh or peal of laughter was an act of pollution. Did the science really support such an extreme measure?

Yes, said Pomerance—the Charney report did *exactly* that. He was beginning to lose his patience, his civility, his stamina. "Now, if everybody wants to sit around and wait until the world warms up more than it has warmed up since there have been humans around—fine. But I would like to have a shot at avoiding it."

Most everybody else seemed content to sit around. The meeting was turning into the same meeting he'd had so often on Capitol Hill. Political appointees confused uncertainty around the margins of the issue (whether warming would be 3 or 4 degrees Celsius in fifty or seventy-five years) for uncertainty about the severity of the problem. As Gordon MacDonald liked to say, carbon dioxide in the atmosphere would cause temperatures to rise; the only question was when. The lag between the emission of a greenhouse gas and the warming it produced could last several decades. It was like adding an extra blanket on a mild night: it took a few minutes before you started to sweat.

Yet Slade, the director of the Energy Department's carbon dioxide program, considered the lag a saving grace. If changes did not occur for another decade or more, he said, those in the room couldn't be blamed for failing to prevent them. So what was the problem?

Pomerance could take it no longer.

"*You're* the problem," he said. Because of the lag between cause and effect, humankind would not likely detect hard evidence of warming until it was too late. The lag would doom them. "The U.S. has to do something to gain some credibility."

"So it *is* a moral stand," replied Slade, as if seizing an advantage.

"Call it whatever." Besides, Pomerance added, they didn't have to ban coal tomorrow. A pair of modest steps could be taken immediately to show the world that the United States was serious: the implementation of a carbon tax and increased investment in renewable energy. Then the U.S. could organize an international summit on climate change. This was his closing plea to the group. The next day, they would have to draft policy proposals.

But when the group reconvened after breakfast on Halloween morning, they immediately became stuck on a sentence in their prefatory paragraph declaring that climatic changes were "likely to occur."

"*Will occur*," proposed Laurmann, the Stanford engineer.

"What about the words *highly likely to occur*?" asked Scoville.

"*Almost sure*," said David Rose, the nuclear engineer from MIT.

"*Almost surely*," said Laurmann.

"Perhaps you can use *will occur*," said another, "and quantify the changes."

"When they start to quantify the changes," said Laurmann, "I can't make such a statement."

"*Changes of an undetermined—*"

"*Changes as yet of a little-understood nature*?"

"*Highly or extremely likely to occur*," said Pomerance.

"*Almost surely to occur*?"

"No," said Pomerance.

"I would like to make one statement," said Annemarie Crocetti, a public health scholar who sat on the National Commission on Air Quality. She had barely spoken all week. "I have noticed that very often when we as scientists are cautious in our statements, everybody else misses the point, because they don't understand our qualifications."

"As a nonscientist," said Tom McPherson, the Florida legislator, "I really concur."

"In 1807," said John Perry, "they didn't have a winter in Switzerland."

Some members of the group exchanged looks.

"There were flowers all winter," said Perry. "I was just reading a book on it. So it is very difficult to make these unequivocal statements that give the world the impression that a group of scientists has sat down and said, you know, we are all going to fry. I, for one, find myself very uncomfortable with some of the language."

"The point," said Scoville, "is that there should be a real push to get them to start making decisions and to really move."

Yet these two dozen experts, who shared the same scientific understanding and had made a commitment to Congress, could not draft a single paragraph. Hours passed in a hell of fruitless negotiation, self-defeating proposals, and impulsive speechifying. Pomerance and Scoville pushed to include a statement calling for the United States to "sharply accelerate international dialogue," but even that was sunk by objections and caveats.

"It is very emotional," said Crocetti, succumbing to her frustration. "What we have asked is to get people from different disciplines to come together and tell us what you agree on and what your problems are. And you have only made vague statements—"

She was interrupted by Waltz, the economist, who wanted simply to note that climate change would have profound effects. "Now, I don't want to go on and elaborate on what I mean by that," he said. "But it is profound."

Crocetti waited until he exhausted himself before resuming in a calm voice. "All I am asking you to say is: 'We got ourselves a bunch of experts and, by God, they all endorse this point of view and think it is very important. They have disagreements about the details of this and that, but they feel that it behooves us to intervene at this point and try to prevent it.'"

They never got to policy proposals. They never got to the second paragraph. The final statement was signed by only the moderator, who phrased it more weakly than the declaration calling for the workshop in the first place. "The guide I would suggest," wrote Jorling, "is whether we know enough *not* to recommend changes in existing policy."

But by that time Rafe Pomerance had already left for Washington. He had seen enough. A consensus-based strategy would not work—could not work—without American leadership. And the United States wouldn't act unless a strong leader persuaded it to do so, someone who could speak with authority about the science, demand action from those in power, and hazard everything in pursuit of justice. Pomerance knew he wasn't that person; he was an organizer, a strategist, a fixer—which was to say that he was an optimist, a romantic even. His job was to build a movement. And every movement, even one backed by widespread consensus, needed a hero. He just had to find one.

7.

A Deluge Most Unnatural

November 1980–September 1981

The meeting ended Friday morning. On Tuesday, four days later, Ronald Reagan was elected president. And Rafe Pomerance found himself wondering whether what had seemed to be a beginning had actually been the end.

In the following months, Reagan floated plans to close the Energy Department, increase coal production on federal land, and deregulate surface coal mining. He appointed James Watt, the president of a legal firm that fought to open public lands to mining and drilling, to run the Interior Department. The president of the National Coal Association pronounced himself "deliriously happy." After some debate about whether to terminate the EPA, Reagan relented and did the next best thing, appointing as administrator Anne Gorsuch, an anti-regulation zealot who proceeded to cut the agency's staff and budget by a quarter. In the midst of this carnage, the Council on Environmental Quality submitted a report to the president warning that fossil fuels could "permanently and disastrously" alter Earth's atmosphere, leading to "a warming of the Earth, possibly with very serious effects." It urged the government to give high priority to the greenhouse effect in national energy policy and to establish a maximum level of carbon dioxide in the global atmosphere. Reagan declined to act on his council's advice. Instead he considered eliminating the council.

At the Pink Palace, Anthony Scoville had said that the problem was not atmospheric but political. That was only half right, Pomerance thought. For behind every political problem, there lay a publicity problem. And the climate crisis had a publicity nightmare. The Florida meeting had failed to articulate a coherent statement, let alone legislation, and now everything was going backward. Even Pomerance himself couldn't devote much time to climate change; Friends of the Earth was busier than ever. The campaigns to defeat the nominations of James Watt and Anne Gorsuch were just the beginning; they were joined by desperate efforts to block mining in wilderness areas, uphold the Clean Air Act's standards for air pollutants, and preserve funding for renewable energy (Reagan "has declared open war on solar energy," said the director of the nation's lead solar-energy research agency, after he was asked to resign). After undoing the environmental achievements of Jimmy Carter, Reagan seemed determined to undo those of Richard Nixon, Lyndon Johnson, John F. Kennedy, and, if he could get away with it, Theodore Roosevelt.

The violence of Reagan's attack alarmed even some members of his own party. Senator Robert Stafford, a Vermont Republican and chairman of the committee that held confirmation hearings on Gorsuch, took the unusual step of lecturing her from the dais about her moral obligation to protect the nation's air and water. Watt's plan to open the waters off California for oil drilling was denounced by the state's Republican senator, and Reagan's proposal to eliminate the position of science adviser was roundly derided by the committee of scientists and engineers who advised him during his presidential campaign. When Reagan threatened to terminate his Council on Environmental Quality, its acting chairman, Malcolm Forbes Baldwin, wrote to the vice president and the White House chief of staff, begging them to ask the president to reconsider; in a major speech the same week, "A Conservative's Program for the Environment," Baldwin argued that it was "time for today's conservatives explicitly to embrace environmentalism." It was not only good sense. It was good business. What could be more conservative than an efficient use of resources that allowed for a reduction of federal subsidies?

The Charney report meanwhile continued to vibrate at the periphery of public consciousness. Its conclusions were confirmed by major studies from the Aspen Institute, the International Institute for Applied Systems Analysis near Vienna, and the American Association for the Advancement of Science. Every month or so there appeared national headlines summoning apocalypse: "Another Warning on 'Greenhouse Effect,'" "Global Warming Trend 'Beyond Human Experience,'" "Warming Trend Could 'Pit Nation Against Nation.'" *People* magazine published a profile of Gordon MacDonald, in which he was photographed standing on the steps of the Capitol, pointing above his head to the level the water would reach when the polar ice caps melted. "If Gordon MacDonald is wrong, they'll laugh," the article read. "Otherwise, they'll gurgle."

But Pomerance knew that to sustain major coverage, you needed major events. Studies were fine; speeches were good; news conferences were better. Congressional hearings, however, were best. The ritual's theatrical trappings—the legislators holding forth on the dais, their aides decorously passing notes, the witnesses sipping nervously from their water glasses, the audience transfixed in the gallery—offered antagonists, dramatic tension, narrative. You couldn't hold a congressional hearing without a scandal, however, or at least a scientific breakthrough. And two years after the Charney group met at Woods Hole, it seemed there was no more science to break through.

It was with a shiver of optimism, then, that Pomerance read on the front page of *The New York Times* on August 22, 1981, about a forthcoming paper in *Science* by a team of seven NASA scientists. They had found that, in the past century, the world had already begun to warm. Temperatures hadn't yet increased beyond the range of historical averages, but the authors predicted that the warming signal would emerge from the noise of routine weather fluctuations much sooner than previously estimated. Most unusual of all, the paper ended with a policy recommendation: in the coming decades, the authors wrote, human civilization should develop alternative sources of energy; fossil fuels should be used only "as necessary." The lead author was listed as Dr. James Hansen.

Pomerance called Hansen to ask for a meeting. He explained that he wanted to make sure he understood the paper's conclusions. More than that, though, he wanted to understand James Hansen.

At the Goddard Institute, Pomerance entered Hansen's office, maneuvering through some thirty piles of documents arrayed across the floor like the skyscrapers of a model city. On top of each stack, some as high as his waist, lay a scrap of cardboard on which had been scrawled words like *Trace Gases*, *Ocean*, *Jupiter*, *Venus*. Behind a desk supporting another paper metropolis, Pomerance found a quiet, composed man with a heavy brow and implacable green eyes. Hansen's speech was soft, equable, deliberate to the point of halting. He would have no trouble passing for a small-town accountant, an insurance-claims manager, or an actuary. And he did hold all those jobs, only his client was the global atmosphere. Pomerance's political sensitivities sparked. He liked what he saw.

Hansen had no political sensitivities—that was clear. But he did have political problems. In the final days of the Carter administration, NASA's budget was cut and Hansen was informed that the funding for his climate research, about $500,000 a year, would in the future need to come from the Energy Department. The director of its carbon dioxide program, David Slade, assured Hansen in writing that he would be good for it. But now Slade was gone. Reagan had replaced him with Fred Koomanoff, a Bronx native with the brusque manner of a sergeant major and an unconstrained zeal for budget cutting. Koomanoff had already summoned Hansen to Washington to justify his research. Hansen did not feel optimistic.

As Hansen spoke, Pomerance listened and watched. He understood Hansen's basic findings well enough: Earth had been warming since 1880, and the warming would reach an "almost unprecedented magnitude" in the next century, leading to the familiar suite of terrors, like the flooding of a tenth of New Jersey and a quarter of Louisiana and Florida. But what excited Pomerance was Hansen's blunt way of making the complex contingencies of atmospheric science sound simple, even intuitive. Though Hansen was something of a wunderkind—at forty, he was about to be named director of the Goddard Institute—he spoke with the blunt midwestern sincerity that played on Capitol Hill. He presented like a heartland voter, the kind of man interviewed on the evening news about the state of the American dream or photographed in the dying sun against a blurry pastoral landscape in a presidential campaign ad. And unlike most scientists in the field, he was not afraid to follow his research to its policy implications. He was perfect.

"What you have to say needs to be heard," said Pomerance. "Are you willing to be a witness?"

8.

Heroes and Villains

March 1982

Though few people other than Rafe Pomerance seemed to have noticed amid Reagan's environmental blitzkrieg, another hearing on the greenhouse effect had been held several weeks earlier, on July 31, 1981, by an obscure House subcommittee. It was led by Representative James Scheuer, a New York Democrat who lived at sea level on the Rockaway Peninsula, in a neighborhood no more than four blocks wide, sandwiched between two beaches, and a canny thirty-three-year-old Democrat from Tennessee named Albert Gore, Jr.

Gore's climate awakening had come a dozen years earlier as an undergraduate at Harvard, when he took a class taught by Roger Revelle. Chalking Keeling's rising zigzag on the blackboard, Revelle explained that humankind was on the brink of radically transforming the global atmosphere and risked bringing about the collapse of civilization. Gore was stunned: If this was true, why wasn't anyone talking about it? He had no memory of hearing about the issue from his father, a three-term senator from Tennessee who later served as chairman of an Ohio coal company. Once in office, Gore figured that if Revelle gave the same lecture before Congress, his colleagues would be moved to act. Or at least that the hearing would get picked up by one of the three major national news broadcasts.

Gore's hearing was part of a larger campaign he had designed with his staff director, Tom Grumbly. After winning his third term in 1980, Gore was granted his first leadership position, albeit a modest one: chairman of an oversight subcommittee within the Committee on Science and Technology—a subcommittee that he had lobbied to create. Most members of Congress considered the science committee a legislative backwater, if they considered it at all; this made Gore's subcommittee, which had no legislative authority, an afterthought to an afterthought. That, Gore vowed, would change. Environmental and health stories had all the elements of narrative drama: villains, victims, and heroes. In a hearing, you could summon all three, with the chairman serving as narrator, chorus, and moral authority. Gore told Grumbly that he wanted to hold a hearing every week.

It was like storyboarding episodes of a procedural drama. Grumbly gamely assembled a list of subjects that possessed the necessary dramatic elements: a Massachusetts cancer researcher who faked his results, the dangers of excessive salt in the American diet, the disappearance of an airplane on Long Island. All fit Gore's template; all had sizzle.

"What about the greenhouse effect?" asked Gore. The issue was apolitical and could not have higher stakes. It seemed certain to get headlines.

Grumbly demurred. "There are no villains," he said. "Besides, who's your victim?"

"If we don't do something," said Gore, "we're all going to be the victims."

He didn't say: *If we don't do something, we'll be the villains too.*

The Revelle hearing, convened on July 31, 1980, at 9:30 a.m. in a small subcommittee room on the second floor of the Rayburn House Office Building, went as Grumbly had predicted. The urgency of the issue was lost on Gore's senior colleagues, who drifted in and out of the room—and consciousness—while the witnesses testified. There were few people left in the gallery by the time the Brookings economist Lester Lave warned that humankind's profligate exploitation of fossil fuels posed an existential test to human nature. "Carbon dioxide stands as a symbol now of our willingness to confront the future," he said, "and especially of our willingness to consider problems that will not be manifest until after the next election. It will be a sad day when we decide that we just don't have the time or thoughtfulness to address those issues." At the end of this particular sad day, Gore's hearing failed to make the nightly news broadcasts, which chose instead to cover the resolution of the baseball strike, the ongoing budgetary debate, and the national surplus of butter.

A few months later, Gore found another opening. Staff members on the science committee heard that the White House planned to eliminate the Energy Department's carbon dioxide program. If they could put a hearing together quickly enough, they figured they could shame the administration before it could execute its plan. The front-page *Times* article about James Hansen's paper had proved that there was a national audience for the carbon dioxide problem—it just had to be framed correctly. Hansen could occupy the role of hero: a sober consummate scientist who had seen the future and sought to rouse the world to action. A villain was emerging too: Fred Koomanoff, Reagan's new director of the Energy Department's carbon dioxide program, a wolf asked to oversee the henhouse. Both men would testify.

Hansen did not disclose to Gore's staff that, in late November, he received a letter from Koomanoff declining to fund his climate-modeling research. Koomanoff left open the possibility of supporting other carbon dioxide studies, but ultimately the rest of Hansen's funding lapsed and he had to release five employees—half his staff. He doubted that Koomanoff could be moved. But the hearing would give Hansen the chance to appeal directly to the congressional leaders who oversaw Koomanoff's budget.

Hansen flew to Washington to testify on March 25, 1982, performing before a gallery even more thinly populated than it had been at Gore's first greenhouse effect hearing. Gore began by attacking the Reagan administration for cutting funding despite the "broad consensus in the scientific community that the greenhouse effect is a reality." William Carney, a Republican from New York, bemoaning the national economy's reliance on fossil fuels, spoke of the need to use science as the basis for policy. Bob Shamansky, a Democrat from Ohio, objected to the use of the term *greenhouse effect* because he had always enjoyed visiting greenhouses. "Everything," he said, "seems to flourish in there." He suggested that they call it the "microwave oven" effect, "because we are not flourishing too well under this. Apparently we are getting cooked."

A Republican from Pennsylvania, Robert Walker, questioned the logic of holding additional hearings on the subject. Hadn't they heard enough already? "Today I have a sense of déjà vu." In each of the last five years, he said, "we have been told and told and told that there is a problem with the increasing carbon dioxide in the atmosphere. We all accept that fact, and we realize that the potential consequences are certainly major in their impact on mankind." Yet Congress had failed to propose a single law. "How frequently must we confirm the evidence before taking remedial steps?" he asked. "Now is the time," he said. "The research is clear."

Gore disagreed: they needed a higher degree of certainty, he believed, to convince a majority of Congress to restrict the use of fossil fuels. The reforms required were of such magnitude and sweep that they "would challenge the political will of our civilization." Nothing less than transformation of the American economy would do it.

Yet the experts invited by Gore agreed with Walker: the science was plenty certain enough. Melvin Calvin, a University of California, Berkeley, chemist who in 1961 had won the Nobel Prize for his work on the carbon cycle, said that it was useless to wait for stronger evidence of warming. "You cannot do a thing about it when the signals are so big that they come out of the noise," he said. "You have to look for early warning signs."

Hansen's job was to share the warning signs, to translate the data into plain English. Not all his work, he explained, involved complex computer models. He also went to libraries. By analyzing records from hundreds of weather stations, he found that the surface temperature of the planet had already increased four-tenths of a degree Celsius. Data from several hundred tide-gauge stations showed that the oceans had risen four inches in the previous century. By comparing old shipping records with current satellite data, he had found that since the 1930s Antarctica had already lost a band of ice, 180 miles in width, rimming the continent. Most disturbing of all, century-old glass astronomy plates had revealed a new problem: some of the more obscure greenhouse gases—especially chlorofluorocarbons (CFCs), a class of synthetic substances used in refrigerators and spray cans—had proliferated wildly in recent years. "We may already have in the pipeline a larger amount of climate change than people generally realize," Hansen told the nearly empty room.

Gore asked when the planet would reach a point of no return—a "trigger point," after which temperatures would spike. "I want to know," he said, "whether I am going to face it or my kids are going to face it."

"Your kids are likely to face it," said Calvin. "I don't know whether you will or not. You look pretty young."

It occurred to Hansen that this was the only political question that mattered: How long until the worst began? It was not a question on which geophysicists expended much effort; the difference between five and fifty years in the future was negligible in geologic time. Politicians were capable of thinking only in terms of electoral time: six years, four years, two years. When it came to the carbon problem, however, the two time schemes were converging.

"Within ten or twenty years," said Hansen at last, "we will see climate changes which are clearly larger than the natural variability."

Representative Scheuer wanted to make sure he understood this correctly. No one else had predicted that the signal would emerge so soon. "If it were one or two degrees per century," he said, "that would be within the range of human adaptability. But we are pushing beyond the range of human adaptability?"

"Yes," said Hansen.

How soon, asked Scheuer, would they need to change the national model of energy production?

Hansen hesitated—it wasn't a scientific question. But he couldn't help himself. He had been irritated, during the hearing, by all the ludicrous talk about the possibility of offsetting emissions by planting more trees. False hopes were worse than no hope at all. They gutted any ambition to develop adequate solutions.

"That time," said Hansen finally, "is very soon."

"My opinion is that it is past," said Calvin, but he was not heard because he spoke from his seat. Scheuer directed him to speak into the microphone. The Nobel laureate rose from his seat and walked to the witness table. He grabbed the microphone.

"It is already later," said Calvin, "than you think."

9.

The Direction of an Impending Catastrophe

1982

Gore considered the hearing an unequivocal success. That night Dan Rather devoted three minutes of *CBS Evening News* to the greenhouse effect. A correspondent explained that temperatures had increased over the previous century, great sheets of pack ice in Antarctica were rapidly melting, the seas were rising; Melvin Calvin announced that "the trend is all in the direction of an impending catastrophe"; and Gore, in a brief appearance, mocked Reagan for his shortsightedness. Later Gore could take credit for protecting the Energy Department's carbon dioxide program, which in the end was largely preserved.

But Hansen's funding was not restored. He wondered whether he had been doomed by his testimony or by his conclusion, in the *Science* paper, that full exploitation of coal resources—a stated goal of Reagan's energy policy—was "undesirable." Whatever the cause, he found himself alone. It was the scenario he had feared, the scenario he had fought for a year to avoid. He felt that he was the victim of some political shadow play whose lineaments he could only dimly perceive. He knew he had done nothing wrong—he had only done diligent research and reported his findings, first to his peers, then to the American people. But now it felt as if he were being punished for it.

Anniek did not entirely share his frustration. With Jim's professional duties reduced, the life of their family came to feel fuller. He began to leave the office at five o'clock to coach his son's baseball team, his daughter's basketball team. (He was a patient, committed coach, detail-oriented, if a touch too competitive for Anniek's liking.) The Hansens moved to a large suburban home in Ridgewood, New Jersey, and expanded it further; their dining room table seated sixteen. Several New York Yankees lived in the neighborhood and on weekends could be seen playing catch with their children in their backyards. At dinner Jim never spoke about work, only about his children's teams and their fortunes—or the Yankees and their fortunes. He didn't even complain about the commute from New Jersey. But Anniek sensed that he was keeping to himself his inner musings—whether he would be able to secure federal funding for his climate experiments, whether the Goddard Institute would be forced to move its office to Maryland to cut costs, whether his future would even be at NASA.

Hansen wondered whether there might be another way

forward. Not long after his modeling research funding ran out, a major symposium he was helping to organize at Columbia University's Lamont-Doherty campus began to receive overtures from a funding partner far wealthier and less ideologically blinkered than the Reagan administration: Exxon. Following Henry Shaw's recommendation to establish credibility ahead of any future legislative battles, Exxon had begun to spend conspicuously on global warming research. It donated tens of thousands of dollars to some of the most prominent research programs, including an international effort coordinated by the United Nations and one at Woods Hole led by the ecologist George Woodwell, who had been calling for major climate policy since the mid-seventies. Now Shaw offered to fund Hansen's symposium at Lamont-Doherty.

As an indication of the seriousness with which Exxon took the issue, Shaw sent in his place his boss, Edward David, Jr., the president of the research division, Nixon's former science adviser, and a member of Reagan's science advisory panel during the transition period. Hansen was glad for the support. He figured that Exxon's contributions might go well beyond picking up the tab for travel expenses, lodging, and the prime rib dinner for the symposium's two dozen attendees at the colonial-style Clinton Inn in Tenafly, the only decent hotel within ten miles of campus. As a gesture of his appreciation, Hansen invited David to give the keynote address.

There were moments in David's speech ("Inventing the Future: Energy and the CO_2 'Greenhouse' Effect") that seemed to channel Rafe Pomerance. He began by attacking capitalism's blind faith in the free market, a doctrine that was "less than satisfying" when it came to environmental protection. Moral considerations, he argued, were necessary too. He pledged that Exxon would revise its corporate strategy to account for climate change, even if it was not "fashionable" to do so. Exxon tended to look no further than twenty years in the future when making business decisions, David explained, but that wouldn't do when it came to the carbon dioxide problem; they needed to consider at least a fifty-year window. In the coming years a great transition from fossil fuels to renewable energy would be necessary, and indeed it had already begun. As Exxon had made deep investments in nuclear and solar technology, David was "generally upbeat" that Exxon would help save the world from the perils of climate change.

Hansen had reason to feel upbeat himself. The frustration of losing his funding had begun to lift, replaced by a dawning sense of relief. It was better, he decided, to be on his own. Without funding, he wouldn't have to follow orders. He wouldn't have to court politicians or bureaucrats, and he wouldn't have to fear reprisals. He was free to follow the research wherever it led. He was free to say anything. And if the world's largest oil and gas company dedicated itself to establishing a new national energy model, the White House would not stand in its way. The Reagan administration was hostile to change from within its ranks. But it couldn't be hostile to Exxon.

It seemed that something was beginning to turn. With the carbon dioxide problem, as with its other thuggish assaults on environmental policy, the administration had alienated many of its own loyalists. But the early demonstrations of autocratic force had retreated into compromise and deferral. By the end of 1982, multiple congressional committees had begun investigating Anne Gorsuch for her indifference to enforcing the cleanup of Superfund sites, and the House voted to hold her in contempt; congressional Republicans turned on James Watt after he eliminated thousands of acres of land from consideration for wilderness designation. Both cabinet members would resign within the year.

The carbon dioxide issue was beginning to trouble the public consciousness—Hansen's own findings had become front-page news, after all. What started as a scientific story was turning into a political story. This prospect would have alarmed Hansen just a couple of years earlier; it still made him uneasy. But he was beginning to understand that politics offered freedoms that the rigors of the scientific ethic denied. The political realm was itself a kind of Mirror World, a parallel reality that, however crudely, mimicked our own. It shared many of the same fundamental laws, like the laws of gravity and inertia and publicity. And if you applied enough pressure, the Mirror World of politics could be sped forward to reveal a new future. Hansen was beginning to understand that too.

Part II
Bad Science Fiction
1983-1988

James Hansen testifying before the Senate on June 23, 1988

10. Caution Not Panic 1983-1984

By the end of 1984, Rafe Pomerance was unemployed, recently recovered from a bout of tachycardia, kicking around a drafty farmhouse in the woods of West Virginia, wondering what he would do with the rest of his life. He knew, from tired experience, that politics moved not in a straight line, but jaggedly, like the Keeling curve—a slow progression interrupted by sharp seasonal declines. But he had entered an especially long, dark winter. The climate issue, after several years of swift progress—during which a question considered esoteric even within the scientific community rose nearly to the level of action, the level at which Republican congressmen made statements like "It is up to us now to summon the political will"—had curdled and died. And all because of a single, lethal report that had done nothing to change the state of climate science but had transformed the state of climate politics.

After the publication of the Charney report in 1979, Jimmy Carter had directed the National Academy of Sciences to prepare a comprehensive, $1 million analysis of the carbon dioxide problem: a Warren Commission for the greenhouse effect. A team of scientist-dignitaries—among them Roger Revelle; the Princeton modeler Syukuro Manabe; veterans of the Charney group and the Manhattan Project; and three future Nobel laureates: the Yale economist William Nordhaus, the astrophysicist William Alfred Fowler, and the Harvard political economist Thomas Schelling, the leading intellectual architect of cold war game theory—were asked to review the literature, evaluate the consequences of global warming for the world order, and propose remedies. Then Reagan won the White House.

For the next three years, as the commission continued its work—drawing upon the contributions of about seventy experts from the fields of atmospheric chemistry, marine biology, geology, astronomy, public health, economics, and political science—the forthcoming report served as the Reagan administration's answer to every question on the subject. There could be no climate policy, Fred Koomanoff and his associates said, until the Academy ruled. In the Mirror World of the Reagan administration, the warming problem hadn't been abandoned at all. A careful, comprehensive solution was being devised. Everyone just had to wait for the Academy's elders to explain what it was.

The commission finally announced its findings on October 19, 1983, in the only setting commensurate with its self-regard: a formal gala, preceded by cocktails and dinner in the Academy's cruciform Great Hall, a secular Sistine Chapel, with vaulted ceilings soaring to a dome painted as the sun. An inscription encircling the sun honored science as the "pilot of industry," and the Academy had invited to the ceremony the nation's foremost pilots of industry, among them a cadre of vice presidents from Peabody Coal, General Motors, and the Synthetic Fuels Corporation. Exxon's Walter Eckelmann, one of the commission's advisers, was there, as was Andrew Callegari, the head of Exxon's carbon dioxide research program. They were eager to learn how the United States planned to act so that they could prepare for the inevitable policy debates. Rafe Pomerance was eager too. But he wasn't invited.

He did manage, however, to secure a spot at the crowded press briefing earlier that day, where he'd snatched up a copy of the 496-page report, *Changing Climate*, and scanned its contents. There had been other blue-ribbon studies in the past four years—by the National Resource Council, the National Climate Program, the World Meteorological Organization, the Australian Academy of Science, and the Energy Department's multifarious carbon dioxide program—all of which had reached the same basic conclusions as the Charney report. And just that week, the EPA had published its own major assessment, *Can We Delay a Greenhouse Warming?* (The EPA's answer, which ran a grim two hundred pages, could be reduced to a word: nope.) But no institutional body had dedicated nearly as much money, time, or expertise to the task as the Academy. The scope of *Changing Climate* was impressive, with its various commissions on agriculture and social policy, its subchapters with titles both specific and grand: "The Colorado River" and "Weeds," "The Deep Circulation" and "The Time Dimension." Nevertheless, as Pomerance flipped through its pages, he could see that it offered no significant new findings. "We are deeply concerned about environmental changes of this magnitude," read the executive summary. "We may get into trouble in ways that we have barely imagined."

The authors did try to imagine some of them: an ice-free Arctic, for instance, and Boston sinking into its harbor, with Beacon Hill surfacing as an island two miles off the coast. There was speculation about political revolution, trade wars, and a long quotation from *A Distant Mirror*, the medieval history written by Barbara Tuchman, Pomerance's aunt, describing how climatic changes in the fourteenth century led to "people eating their own children" and "feeding on hanged bodies taken down from the gibbet." The committee's chairman, William Nierenberg—a Jason, a presidential adviser, and Roger Revelle's successor as the director of the Scripps Institution of Oceanography at UC San Diego, the nation's preeminent oceanographic center—argued in his preface that action had to be taken immediately, before all the details could be known with certainty, or else it would be too late.

That's what Nierenberg wrote in *Changing Climate*. But it's not what Nierenberg and other august members of the central committee emphasized in the press interviews that followed. They argued the opposite: there was no urgent need for action. Nierenberg warned that the public should not entertain the most "extreme negative speculations" about climate change (despite the fact that many of those speculations appeared in his report). Though *Changing Climate* urged an accelerated transition to renewable fuels, noting that it would take thousands of years for the atmosphere to recover from the damage of the last century, Nierenberg recommended "caution, not panic." It was a serious problem, granted, but "if it goes the way we think, it will be manageable in the next hundred or so years." Better to wait and see. Better to bet on American ingenuity to save the day. Major, immediate interventions in national energy policy might end up being more expensive, and less effective, than actions taken decades in the future, after more was understood about the economic consequences of a warmer planet. Yes, the climate would change, mostly for the worst, but future generations would be better equipped to change with it.

This line was echoed by Roger Revelle himself. "We're flashing a yellow light but not a red light," he told reporters. "It's not an unmitigated disaster by any means. It's just a change." A third prominent member of the central committee, Joseph Smagorinsky, a pioneering climate modeler who helped found the lab at Princeton where Syukuro Manabe ran his global warming models, openly denigrated the "unnecessarily alarmist" EPA report. He reserved particular contempt for its use of projections, extending more than a century into the future, that assumed humanity would continue to consume increasing quantities of fossil fuels. "If you do that," said Smagorinsky, "you get fantastic numbers." Nierenberg called the EPA report "a badly done thing." But in the end the authors of the two reports endorsed the same response, with the EPA arguing that it was already too late to avoid the worst, and the Academy that it was too early. Both suggested that adaptation was the only possible outcome. As Thomas Schelling put it in the pages of *Changing Climate*, echoing Klaus Meyer-Abich, no regulatory policy could possibly succeed, "so climate change is what we should expect." It was a self-fulfilling prophecy.

While Pomerance tried to absorb the commission's appeasements, he glanced, baffled, around the briefing room. The reporters and staff members listened politely to the presentation and took dutiful notes, as at any technical briefing. The officials in the room who knew Nierenberg were not surprised by his conclusions: he was an optimist by training and experience, superior and pedantic, a devout believer in the doctrine of American exceptionalism, a member of the Curia Regis of scientists who had helped guide every president since Franklin D. Roosevelt through economic despair, the nuclear age, and the cold war. These scientists, many of whom had contributed to *Changing Climate*, had helped to restore the plains after the Dust Bowl, invented the bomb and won World War II, developed the booming aerospace and computer industries. They had solved every existential crisis the nation had faced over the previous generations. Surely they would not be daunted by an excess of a gas that human beings exhaled with each breath. Nierenberg had served on Reagan's science and technology task force during the transition period—he was passed over for the job of science adviser—and his political sensibility reflected all the ardor of his party: sanguine about the saving graces of market forces, skeptical of government intervention.

Pomerance, having come of age during the Vietnam War and the birth of the environmental movement, shared none of this procrustean faith in American ingenuity. He feared the dark undertow of rapid industrial advancement, the way that each new technological superpower carried within it unintended consequences that, if unchecked, eroded the foundations of society. Technology had not solved the air and water crises of the 1970s. Activism and organization, forcing robust government regulation, had. Hearing Nierenberg's equivocations, he shook his head, rolled his eyes, groaned. He felt that he was the only sane person in a briefing room gone mad. How could a committee that included Roger Revelle, George Woodwell, and William Nordhaus counsel doing nothing? Revelle had been warning presidents since the Eisenhower administration; Woodwell had drafted the 1970 letter to the Carter White House, signed by Revelle, Gordon MacDonald, and Charles David Keeling, declaring that the window for reducing dependence on fossil fuels "is fast passing"; and Nordhaus, in the very pages of *Changing Climate*, made the case for a carbon tax—an argument far more audacious than the EPA's ambivalent position. To conclude, from all this, that no action should be taken was not only insane; it was *wrong*. Someone in the next row told Pomerance to calm down.

Nierenberg's press release for *Changing Climate*, being one-hundredth the length of the actual assessment, received one hundred times the press coverage. As *The Wall Street Journal* put it, in a line echoed by trade publications across the nation: "A panel of top scientists has some advice for people worried about the much-publicized warming of the

Earth's climate: You can cope." The effusiveness of the Academy scientists' reassurances invited derision. On *CBS Evening News*, Dan Rather said they had given "a cold shoulder" to the EPA assessment published earlier that week. *The Washington Post* described the two reports, taken together, as "clarion calls to inaction."

The greatest blow, however, came from *The New York Times*, which published its most prominent piece on global warming to date under the headline "Haste on Global Warming Trend Is Opposed." Although the paper included an excerpt from *Changing Climate* that detailed some of its gloomier predictions, the article itself gave the greatest weight to a statement, heavily workshopped by White House senior staff, credited to George Keyworth II, Reagan's science adviser. Keyworth used Nierenberg's optimism as reason to discount the EPA's "unwarranted" report and warned against taking any "near-term corrective action." In case the administration's position was not entirely clear, Keyworth added for emphasis, "There are no actions recommended other than continued research."

In the following weeks, press coverage withered and the industry tuned out. The American Petroleum Institute disbanded its CO_2 task force; Exxon ended its carbon dioxide program. In a presentation at an industry conference, Henry Shaw cited *Changing Climate* as evidence that "the general consensus is that society has sufficient time to technologically adapt to a CO_2 greenhouse effect." If the Academy had concluded that emissions regulations were not a serious option, why should Exxon make a fuss? Edward David, Jr., two years after boasting of Exxon's commitment to transforming global energy policy, told *Science* that the corporation had reconsidered: "Exxon has reverted to being mainly a supplier of conventional hydrocarbon fuels—petroleum products, natural gas, and steam coal." He spoke of "going back to the fundamentals."

In a frantic effort to resuscitate the issue, Pomerance helped draw together another hearing before Al Gore's subcommittee, on February 28, 1984, to review the findings of the two federal reports. Pomerance asked to testify himself. It didn't come naturally—he preferred to let those with greater expertise or celebrity speak for him. He had a talent for convincing people that his idea was theirs, that they'd thought of it themselves, had always wanted to act on it, and just hadn't had the chance until now. But there was no time for that approach anymore. He was desperate. Gore shared his desperation. He spoke with an urgency that at times flagged into resignation, likening the greenhouse effect to "a bad science fiction novel," its ramifications so inconceivable—Manhattan as balmy as Palm Beach, Kansas posing as central Mexico—as to make measured debate seem frivolous.

Gore had summoned as witnesses the figures he deemed most capable of producing clips for the nightly news. Carl Sagan called it "the height of irresponsibility" to alter the global environment without grappling with the consequences. Wallace Broecker presented an ominous new prophecy: deeply buried ice, recently excavated from Antarctica and Greenland, revealed that radical climate changes did not occur gradually, as previously assumed, but in sudden wild "jumps" that reorganized the circulation of the oceans and could lead to cataclysms worse than anything yet imagined. John Hoffman, now the director of the EPA's carbon dioxide research office, calculated that a sea level rise of a single foot might trim as much as two hundred feet off the Atlantic Coast. Appearing in defense of *Changing Climate* was Thomas Malone, a genial, highly regarded member of the Academy who had written the report's foreword.

Malone had been one of the first American scientists to speak out about the carbon dioxide issue; in 1966 he oversaw an Academy report, chaired by Gordon MacDonald, on humankind's ability to change the global climate, and later that year he testified before Congress that the danger of global warming "is something we must resolve in a matter of decades." Yet now, nearly two decades later, he repeated Nierenberg's reassurances, urging Congress not to take "premature" action. Pomerance, seated beside Malone, could not disguise his reaction.

"I know you're itching to say something, Mr. Pomerance," said Gore.

"It is time to act," said Pomerance. "We know what to do. The evidence is in. The problem is as serious as exists. People talk about not leaving this to their grandchildren. I'm concerned about leaving this to my *children*." The whole premise of the conversation was confused, he said. It was ridiculous to wait for scientists to demand action. They already agreed on the basic facts, after all. Why not place the burden on the energy industries? Ask them to prove that burning fossil fuels was benign. The Academy's caution was inane, while the EPA, in its pessimism, failed to emphasize that it was still possible to avoid the worst-case scenarios. But action had to come immediately. Even if coal were banned tomorrow, it would take years, if not decades, to phase out. Civilization was a locomotive hurtling over a rickety bridge toward an abyss. The track light was blinking red, the boom barrier was descending, but the train had too much momentum—

"You are the ones who are going to have to make that decision," he told Gore. "Don't rely on the scientists. It's not their job."

What could Congress do? Pomerance had come with an action plan, which he entered into the record: prepare for the climatic changes that were inevitable; fund more research; make conservation the highest consideration in all energy policy; and abolish the federal synthetic-fuels initiative. These measures would have the added benefits of reducing acid rain, increasing energy security, promoting public health, and saving money. "This issue is so big," said Pomerance, "yet the attitude that is being taken is so relaxed. I mean, it strikes one as a bit incredible."

He took a breath before concluding. "The major missing element in all this is leadership," he said. "It needs to come from the political community."

Those remaining in the hearing room turned to the congressman from Tennessee, who, in the previous hearing, had spoken of the challenge posed by the warming problem to the political will of civilization. In order to justify major policy, he said, the nation needed a high degree of certainty about the science; yet the potential consequences were too horrific to delay much longer. "It is a hard one," he said. "It is really hard." The hearing adjourned a few minutes later.

Not long after the hearing, Pomerance resigned from Friends of the Earth. He credited various factors: he had struggled with the politics of managing a staff and a board; he'd had disagreements with David Brower, the organization's leader in influence if no longer in title; and the environmental movement from which Friends of the Earth had emerged in the early seventies was in crisis. It lacked a unifying cause. Climate change, Pomerance believed, could be that cause. But its insubstantiality made it difficult to rally the older activists, whose strategic model relied on protests at sites of horrific degradation—Love Canal, Hetch Hetchy, Three Mile Island. How did you stage a protest when the toxic waste dump was the entire planet or, worse, its invisible atmosphere?

Observing her husband, Lenore Pomerance was reminded of an old *Philadelphia Bulletin* ad campaign: "In Philadelphia—nearly everybody reads the *Bulletin*." In one of the spots, all the passengers on a crowded commuter train bury their faces in their newspapers, except for one man, who stares out the window into the distance. Here Rafe, the loner, was staring down the world's largest problem, while everyone else had their heads in their laps, lost in the news of the day. Pomerance didn't like talking about his work and acted cheerful at home, fooling his kids. But he couldn't fool Lenore. And he couldn't fool his nervous system. Near the end of his tenure at Friends of the Earth, a doctor diagnosed him with an abnormally high heart rate.

Pomerance planned to take a couple of months to reflect on what he should do next. Two months stretched to a year. He brooded; he checked out. He spent weeks at a time at an old farmhouse that he and Lenore owned in West Virginia, near Seneca Rocks. When they had bought it in the early seventies, it had a wood-burning stove and lacked running water. To make a phone call on a private line, you had to drive to the operator's house and hope she was in. Pomerance sat in the cold house and he thought.

The winter returned him to his childhood. He grew up in Cos Cob, Connecticut, on part of an estate purchased by his grandfather, a banker and conservationist named Maurice Wertheim, in 1912. A short walk from Pomerance's home brought him to a pond on which his mother taught him to ice-skate. He remembered the muffled hush of twilight, the snow dusting the ice, the ghostly clearing surrounded by a wood darker than the night. His house had been designed by his father, an architect whose glass-enveloped structures mocked the vanity of humankind's efforts to improve on nature; the broad expansive windows invited the elements inside, the trees and the ice and, in the rattling of the broad panes, the wind. Winter, Pomerance believed, was part of his soul. When he imagined the future, he worried about the loss of ice, the loss of spiky Connecticut January mornings. He worried about the loss of some irreplaceable part of himself.

He wanted to recommit himself to the fight but he couldn't figure out how. During the past five years he had tried every tactic that had sustained the environmental wars of the seventies. Nothing had worked. The carbon dioxide issue had fallen off the national agenda. If scientists, the intelligence community, Congress, and the national press could not force action, who could? He didn't see what was left for him to do. He didn't see what was left for anyone to do. He didn't see that the answer was at that moment floating over his head, about ten miles above his West Virginia farmhouse, just beyond the highest clouds in the sky.

11.

The World of Action

1985

It was as if, without warning, the sky opened and the sun burst through in all its irradiating, blinding fury. The mental image was of a pin stuck through a balloon, a chink in an eggshell, a crack in the ceiling—Armageddon descending from above. It was a sudden global emergency: there was a hole in the ozone layer.

The claxon was cranked by a team of British government scientists, until then little known in the field, who made regular visits to research stations in Antarctica—one on the Argentine Islands, the other on a sheet of ice drifting out into the sea at the rate of a quarter mile per year. At each site, the scientists had set up a machine invented in the 1920s called the Dobson spectrophotometer, which resembled a large slide projector turned with its eye staring straight up. After several years of results so alarming that they had disbelieved their own evidence, the British scientists finally reported their discovery in the May 1985 issue of *Nature*. "The spring values of total O_3 in Antarctica have now fallen considerably," the abstract read. But by the time the news filtered into national headlines and television broadcasts several months later, it had transfigured into something horrific: a substantial increase in skin cancer, a sharp decline in the global agricultural yield, and the mass death of fish larvae, one of the first links in the marine food chain. Later came fears of atrophied immune systems and blindness; one activist likened the ozone hole to "AIDS from the sky."

The urgency of the alarm seemed to have everything to do with the phrase "a hole in the ozone layer," which, charitably put, was a mixed metaphor. For there was no hole, and there was no layer. Ozone, which shielded Earth from ultraviolet radiation, was distributed throughout the atmosphere, settling mostly in the middle stratosphere and never in a concentration higher than fifteen parts per million. As for the "hole"—while the levels of ozone over Antarctica had declined drastically, the depletion was a temporary phenomenon, lasting about two months a year. In satellite images colorized to show ozone density, however, the darker region appeared to depict a void. When F. Sherwood Rowland, one of the chemists who identified the problem in 1974, spoke of the "ozone hole" in a university slide lecture in November 1985, the crisis found its catchphrase. *The New York Times* borrowed it for an article that same day, and though scientific journals initially refused to use the term, within a year it was unavoidable. The ozone crisis had its signal, which was also a symbol: a hole.

It was already understood, thanks to the work of Rowland and his colleague Mario Molina, that the damage was largely caused by the synthetic chlorofluorocarbons used in refrigerators, spray bottles, and plastic foams, which escaped into the stratosphere and vampirized ozone molecules. It was also understood that the ozone problem and the greenhouse gas problem were linked. CFCs were unusually potent greenhouse gases. Though CFCs had been mass-produced only since the 1930s, they were already responsible, by Jim Hansen's calculation, for nearly half of Earth's warming during the 1970s. But nobody was worried about CFCs because of their warming potential. They were worried about going blind.

The United Nations, through two of its intergovernmental agencies—the United Nations Environment Programme (UNEP) and the World Meteorological Organization (WMO)—had in 1977 established the World Plan of Action on the Ozone Layer. In 1985, UNEP established a framework for a global treaty, the Vienna Convention for the Protection of the Ozone Layer. The negotiators failed to agree on any specific CFC regulations in Vienna, but two months later, after the British scientists reported their Antarctic findings, Reagan proposed a reduction in CFC emissions of 95 percent. It was a sudden, stunning reversal. Just several months earlier, the National Resources Defense Council had sued the EPA for failing to propose a single regulation, despite its obligations under the Clean Air Act. Since taking power, the Reagan administration had dutifully parroted the arguments of the Alliance for Responsible CFC Policy, a lobbying group founded in 1980 that represented nearly every U.S. business that had the word *refrigeration* in its name or was involved in the production, manufacture, and consumption of chemicals, plastics, paper goods, and frozen food—five hundred companies in total, from DuPont and the American Petroleum Institute to Mrs. Smith's Frozen Food Company. The alliance hounded the EPA, members of Congress, and Reagan himself with a single message: there was too much uncertainty in the science to justify any further regulation of CFCs. But once the public discovered the "ozone hole," every relevant government agency and every sitting U.S. senator urged the president to endorse the UN's plan for a treaty. When Reagan finally submitted the Vienna Convention to the Senate for ratification, he praised the "leading role" played by the United States, fooling nobody.

Senior members of UNEP and the WMO began to wonder whether they could do for the carbon dioxide problem what they had done for ozone. The organizations had been holding semiannual conferences on global warming since the early 1970s. But in 1985, just several months after the bad news from the Antarctic, at an otherwise sleepy meeting in the Carinthian city of Villach, eighty-nine scientists from twenty-nine countries began to discuss a subject that fell wildly outside their fields of expertise: politics.

Many of the most valuable conversations did not take place during the meeting hours, which were dominated by presentations, but at night, at taverns over glasses of Zweigelt and Blaufränkisch, or on the terraces at the Hotel Post, with their panoramic views of Austria's Alpine foothills. The old guard was well represented—Roger Revelle, Syukuro Manabe, Thomas Malone—and its members had long lost their ability to be surprised by any of the dire predictions. But the newcomers were stunned to learn of the severity of the threat. Many had practical concerns. An Irish hydrology expert asked whether his country should reconsider the location of its dams. A Dutch seacoast engineer questioned the wisdom of rebuilding dikes that had been destroyed by recent floods. And the conference's chairman, James Bruce, an unassuming, pragmatic hydrometeorologist from Ontario, posed a question that unsettled the conference.

Bruce was a minister of the Canadian environmental agency, a position that conferred on him the esteem that his American counterparts had forfeited when Reagan took the White House. Just before leaving for Villach, he had met with provincial dam and hydropower managers. OK, one of them said, you scientists win. You've convinced me that the climate is changing. Well, tell me how it's changing. In twenty years, will the rain be falling somewhere else?

Bruce took this challenge to Villach: Well, gang, you're the experts. What am I supposed to tell him? You guys have been talking about this stuff for years. People are hearing the message and they want practical guidance. So how do we, in the scientific world, begin a dialogue with the world of action?

The world of action. For a room of scientists who prided themselves as belonging to a specialized guild of monkish austerity, this was a startling provocation. On a bus tour of the countryside, commissioned by their Austrian hosts, Bruce sat with Roger Revelle, ignoring the Alps, speaking animatedly about the need for scientists to demand political remedies in times of existential crisis.

Within a couple of days, there were some notable conversions, none more striking than that of Thomas Malone, who just the previous year, speaking on behalf of the National Academy, had urged Congress not to take action on climate change. At the final assembly in Villach, he stood before his peers and atoned. "As a reversal of an opinion I held a year or so ago," said Malone, "I believe it is timely to start on the long, tedious, and sensitive task of framing a convention on greenhouse gases, climate change, and energy." There was some mumbling in the room. Malone was about as reliable a bellwether for the scientific establishment as there was; if he had come around to the notion that scientists should advocate for policy, anyone could join him without the risk of being considered a radical. James Bruce stood up and declared that it was time to begin serious consideration of "the costs and benefits of a radical shift away from fossil fuel consumption"—even if they were only a bunch of geophysicists.

The formal report ratified at Villach contained the most forceful warnings yet issued by a scientific body. Most major economic decisions undertaken by nations, it pointed out, were based on the assumption that past climate conditions were a reliable guide to future conditions. But the future would not resemble the past. Though some warming was inevitable, the scientists acknowledged, the extent of the disaster could be "profoundly affected" by aggressive, coordinated government policies. Fortunately they had established, with the ozone treaty, a new international model to accomplish just that. The balloon could be patched, the eggshell bandaged, the ceiling replastered. There was still time.

12.

The Ozone in October

Fall 1985–Summer 1986

It was fall 1985, and Curtis Moore, a Republican staff member on the Committee on Environment and Public Works, was telling Rafe Pomerance that the greenhouse effect wasn't a problem.

With his last ounce of patience, Pomerance begged to disagree.

Yes, Moore clarified, of course, it was an existential problem—the fate of civilization depended on it, the oceans would boil, all of that. But it wasn't a *political* problem. Know how you could tell? Political problems had solutions. And the climate issue had none. Without a solution—an obvious, attainable one—any policy could only fail. No elected politician desired to come within shouting distance of failure.

So when it came to the dangers of despoiling our planet beyond the range of habitability, most politicians didn't see a problem. Which meant that Pomerance had a very big problem indeed.

He had followed the rapid ascension of the ozone issue with the rueful admiration of a competitor. He was thrilled for its success—however inadvertently, the treaty would serve as the world's first action to delay climate change. But it offered an especially acute challenge for Pomerance, who, after his yearlong hiatus, had become the nation's first, and only, full-time global warming lobbyist. At the suggestion of Gordon MacDonald, Pomerance joined the World Resources Institute, a nonprofit begun by Gus Speth, formerly of Jimmy Carter's Council on Environmental Quality and a founder of the National Resources Defense Council. Unlike Friends of the Earth, WRI was not an activist organization; it occupied the nebulous intersection of politics, international diplomacy, and energy policy; its advisory board was populated by veteran environmentalists as well as executives from Dow Chemical and Exxon. Its mission was expansive enough to allow Pomerance to work without interference on developing policy solutions to global warming. Yet the only thing that anyone on Capitol Hill wanted to talk about was ozone.

That was Curtis Moore's proposal: use ozone to revive carbon dioxide. The ozone hole had a solution—an international treaty, under the auspices of the United Nations, already in negotiation. Why not hitch the milk wagon to the bullet train?

Pomerance was skeptical. The problems were related,

sure: without a reduction in CFC emissions, you didn't have a chance of averting cataclysmic global warming. But it had been difficult enough to explain the carbon issue to politicians and journalists; why complicate the sales pitch? Then again, he didn't see what choice he had. The Republicans controlled the Senate, and Moore was his connection to the Senate's environmental committee.

Moore came through. At his suggestion, Pomerance and Gus Speth met with Senator John Chafee, the Rhode Island Republican who chaired the Senate subcommittee on environmental pollution, and helped persuade him to hold a double-barreled hearing on the twin problems of ozone and carbon dioxide. On June 10 and 11, 1986, F. Sherwood Rowland, NASA's Robert Watson, and Richard Benedick, the administration's lead representative in international CFC negotiations, would discuss ozone; James Hansen, Al Gore, George Woodwell, and Carl Wunsch, a Charney group alumnus, would testify about climate change. As soon as the first witness appeared, Pomerance realized that Moore's instincts had been right. The ozone gang was good.

Robert Watson dimmed the lights in the hearing room. On a flimsy projection screen, he beamed footage with the staticky, low-budget quality of a slasher flick. It showed a bird's-eye view of the Antarctic, partly obscured by spiraling clouds. The footage looked so realistic that Chafee had to ask whether it was an actual satellite image. Watson acknowledged that though *created* by satellite observations, it was, in fact, a simulation, a pictorial representation of the data.

An animation, to be precise. The three-minute time-lapse

video showed every day of October—the month during which the ozone thinned most drastically—for seven consecutive years. (The other months, conveniently, were omitted.) A canny filmmaker had colored the "ozone hole" pink. As the years sped forward, the polar vortex madly gyroscoping, the pink smudge underwent a peristaltic expansion until it obscured most of Antarctica. The smudge turned mauve, representing an even thinner density of ozone, and then the dark purple of a hemorrhaging wound. The data represented in the video wasn't new, but nobody had thought to represent it in this medium. If F. Sherwood Rowland's earlier colorized images were crime-scene photographs, Watson's video was a surveillance camera catching the killer red-handed.

As Pomerance had hoped, fear about the ozone layer guaranteed a paroxysm of press coverage for the hearings. As he had feared, it caused many casual observers to conflate the two crises. One was Peter Jennings, who aired the video on ABC's *World News Tonight*, warning that the ozone hole "could lead to flooding all over the world, also to drought and to famine."

The confusion helped. For the first time since *Changing Climate*, global warming headlines appeared by the dozen. William Nierenberg's "caution, not panic" line was inverted. It was all panic without a hint of caution: "A Dire Forecast for 'Greenhouse' Earth" (the front page of *The Washington Post*); "Scientists Predict Catastrophes in Growing Global Heat Wave" (*Chicago Tribune*); "Swifter Warming of Globe Foreseen" (*The New York Times*). On the second day of the hearing, devoted to global warming, every seat in the gallery was occupied; four men squeezed together on each of the broad windowsills, their legs dangling, like schoolchildren.

Speth and Pomerance had proposed to Chafee that, instead of opening with the familiar call for more research, he demand action. But Chafee went further: he requested that the State Department begin negotiations on a global climate accord with the Soviet Union. It was the kind of proposal that would have been unthinkable even a year earlier, but the ozone melee had established a precedent for international environmental problems: high-level meetings among the world's great powers, followed by a summit meeting to negotiate a framework for a binding treaty.

After three years of backsliding and silence from the federal government, Pomerance was exhilarated to see interest in the issue spike overnight. Not only that: a solution materialized, and a moral argument passionately articulated—by Rhode Island's Republican senator no less. "Ozone depletion and the greenhouse effect can no longer be treated solely as important scientific questions," said Chafee. "They must be seen as critical problems facing the nations of the world, and they are problems that demand solutions."

The old canard about the need for more research was roundly mocked—by Woodwell, by a WRI colleague of Pomerance's named Andrew Maguire, and by Senator George Mitchell, a Democrat from Maine. "Scientists are never one hundred percent certain," testified Theodore Rabb, a Princeton historian. "Even Newton has been proven to be wrong. That notion of total certainty is something too elusive ever to be sought." James Hansen, in his prepared statement, wrote that "evidence confirming the essence of the greenhouse theory is already overwhelming from a scientific point of view." As Pomerance had been saying since 1979, it was past time to act. Only now the argument was so widely accepted that nobody dared object.

The ozone hole, Pomerance realized, alarmed the public because, though it was no more visible than global warming, ordinary people could be made to see it. They could watch it gestate on videotape. Its metaphors were emotionally wrought: instead of summoning a glass building that sheltered plants from chilly weather ("Everything seems to flourish in there"), the hole evoked a violent rending of the firmament, inviting deathly radiation. Newspapers advised greater vigilance in the application of sunscreen. Americans felt that their lives, and the lives of their children, were in danger. An abstract, atmospheric problem had been reduced to the size of the human imagination. It had been made just small enough, and just large enough, to break through.

13.

Atmospheric Scientist, New York, N.Y.

Fall 1987–Spring 1988

Four years after *Changing Climate*, two years after a hole had torn open the sky, and a month after the United States and more than three dozen other nations signed a treaty to limit the use of CFCs, the climate change corps was ready for a party. It had become conventional wisdom that the issue would follow ozone's route of ascent into international law. The head of Reagan's EPA, Lee M. Thomas, said as much the day he signed the Montreal Protocol on Substances That Deplete the Ozone Layer, telling reporters that global warming would likely be the subject of a similar international agreement. The political momentum had flipped. Now that the ozone problem was on the verge of being "fixed," climate issues had once again become a popular excuse for hearings on Capitol Hill—a noncontroversial subject that elicited concern, headlines, and expressions of moral grandstanding and American might. In 1987 alone, there were eight days of climate hearings, in three committees, across both chambers of Congress; Senator Joseph Biden, a Delaware Democrat, had introduced legislation to establish a formal national climate change strategy. And so it was that Jim Hansen found himself on October 27 in the not especially distinguished ballroom of the Quality Inn on New Jersey Avenue, a block from the Capitol, at "Preparing for Climate Change," which was technically a conference but felt more like a wedding.

The mood was inspired by the host: John Topping, big-hearted and curious, with the infectious enthusiasm of an autodidact, was an old-line Rockefeller Republican, a Commerce Department lawyer under Nixon, and an EPA official under Reagan. He had first heard about the climate problem in the halls of the EPA in 1982 and sought out Hansen, who gave him a personal tutorial. Topping was amazed to discover that only seven people out of the EPA's thirteen-thousand-person staff were assigned to work on climate, though he figured it was more important to the long-term security of the nation than every other environmental issue combined. Even sober, steady William Ruckelshaus, Reagan's replacement for Anne Gorsuch as EPA administrator, had delivered a pair of speeches in 1984 acknowledging that a failure to reduce dependence on fossil fuels would lead to "a succession of unexpected and shattering crises" and "threaten all we hold dear." After Topping left the administration, he founded a nonprofit organization, the Climate Institute, to unite scientists, politicians, and businesspeople to solve the problem. He didn't have any difficulty raising $150,000 from BP America, General Electric, and the American Gas Association, contacts from his tenure at the EPA, to fund the conference. His industry friends were intrigued. If a guy like Topping thought this greenhouse business was important, they'd better see what it was all about.

Glancing around the room, Jim Hansen could chart, like an arborist counting rings on a stump, the growth of the climate affair over the decade. Veterans like Gordon MacDonald, George Woodwell, and the environmental biologist Stephen Schneider stood at the center of things. Former and current staff members from the congressional science committees (Tom Grumbly, Curtis Moore, Anthony Scoville) made introductions to the congressmen they advised (John Chafee, George Mitchell, George Brown). Hansen's owlish nemesis Fred Koomanoff was present, as were his counterparts from the Soviet Union and Western Europe. Rafe Pomerance's cranium could be seen above the crowd, but unusually he was surrounded by colleagues from other environmental organizations that, until now, had shown little interest in a diffuse problem with no proven fund-raising record. The party's most conspicuous newcomers, however, the outermost ring, were the oil and gas executives.

It was not entirely surprising to see envoys from Exxon, the Gas Research Institute, and the electrical-grid trade groups, even if they had gone silent since *Changing Climate*. But they were joined by executives from the American Petroleum Institute, which that spring, at the industry's annual world conference in Houston, had invited a leading government scientist to make the case for a transition to renewable energy; the speech was titled "The Reality of the Greenhouse Effect." Even Richard Barnett, the chairman of the Alliance for Responsible CFC Policy, the face of the campaign to defeat an ozone treaty, was there. Barnett's retreat had been humiliating and swift. After DuPont, by far the world's single largest manufacturer of CFCs, realized that it stood to profit from the transition to replacement chemicals and began placing full-page ads in *The New York Times* to announce its support for a phaseout, the alliance abruptly reversed its position, demanding that Reagan sign a treaty as soon as possible. Now Barnett, at the Quality Inn, was claiming to "bask in the glory of the Montreal Protocol" and quoting Robert Frost's "The Road Not Taken" to emphasize his earnest hope for a coalition between industry and environmental activists. There were more than 250 people in the old ballroom, and if the concentric rings had extended any further, they would have needed a larger hotel.

That evening, as a storm spat and coughed outside, Rafe Pomerance gave one of his exhortative speeches urging cooperation among the various factions; John Chafee and Roger Revelle received awards; and introductions were made and business cards exchanged. Not even a presentation by Hansen of his research could sour the mood. The next evening, at a high-spirited dinner party in Topping's town house on Capitol Hill, the oil and gas men joked with the environmentalists, the trade-group representatives chatted up the regulators, and the academics got merrily drunk. Mikhail Budyko, the don of Soviet climatology, settled into an engrossing conversation about global warming with Topping's ten-year-old son. It all seemed like the start of a grand bargain, a virtuous realignment—a solution.

It was because of all this good cheer that Hansen was inclined to shrug off a peculiar series of events that occurred just a week later. He was scheduled to appear at another Senate hearing, this time devoted entirely to climate change. It was called by the Committee on Energy and Natural Resources after Rafe Pomerance and Gordon MacDonald persuaded its chairman, Bennett Johnston, a Democrat from Louisiana, of the issue's significance to the future of the oil and gas industry. Hansen was accustomed to the bureaucratic nuisances that attended testifying before Congress as a federal employee; before a hearing, he had to send his formal statement to NASA headquarters, which forwarded it to the White House's Office of Management and Budget for approval. "Major greenhouse climate changes are a certainty," he had written. "By the 2010s [in every scenario], essentially the entire globe has very substantial warming." Until now the process had appeared perfunctory. But on the Friday evening before his Monday appearance, he was informed that the White House demanded changes to his testimony.

No rationale was provided. Nor could Hansen understand by what authority the White House could censor scientific findings. He told the administrator in NASA's legislative affairs office that he refused to make the changes. If that meant he couldn't testify, so be it.

The NASA administrator had another idea. The Office

of Management and Budget, she explained, had the authority to approve government witnesses. But it couldn't censor a private citizen.

At the hearing on November 9, Hansen was listed as "Atmospheric Scientist, New York, N.Y."—as if he were a crank with a telescope who had stumbled into Dirksen off the street. He was careful to emphasize the absurdity of the situation in his opening remarks, at least to the degree that his midwestern reserve would allow: "Before I begin, I would like to state that although I direct the NASA Goddard Institute for Space Studies, I am appearing here as a private citizen." In the most understated terms available to him—one of the world's foremost authorities on climate modeling, the person responsible for many of the decisive findings about the greenhouse effect in the previous decade—Hansen provided his credentials: "ten years' experience in terrestrial climate studies and more than ten years' experience in the exploration and study of other planetary atmospheres."

Assuming that one of the senators would immediately question this bizarre introduction, Hansen had prepared an elegant response. He planned to say that, although his NASA colleagues endorsed his conclusions, the White House had insisted he utter false statements that would have distorted them. He figured this would lead to an uproar. But no senator thought to ask about his title. So the atmospheric scientist from New York City said nothing else about it.

After the hearing he went to lunch with John Topping, who was stunned to hear of the White House's ham-handed attempt to silence him. "Uh, oh," joked Topping, "Jim is a dangerous man. We're going to have to rally the troops to protect him." The idea that quiet, sober Jim Hansen could be seen as a threat to anyone, let alone national security—well, it was enough to make him burst into laughter.

Still the brush with state censorship stayed with Hansen in the months that followed, long after the exuberance of the Climate Institute gathering had faded. It weighed on him, this desire to silence him, and after some reflection Hansen decided that it wasn't very funny, not at all.

At first he had assumed that it was enough to publish articles in the nation's most prestigious journals about global warming and that federal policymakers, perceiving the gravity of the situation, would spring into action. Then he figured that his statements to Congress, reported in the national press, would do it. It had seemed, at least momentarily, that the energy industry, understanding what was at stake for its future, might lead. But nothing had worked. And now, even after the political triumph of the Montreal Protocol and the bipartisan support of climate policy, he realized there were people at the highest levels of the federal government—within the White House itself—who hoped to prevent so much as an honest reckoning with the nature of the problem. This, it seemed, was a new development: not merely an expression of indifference or caution, but the emergence of an antagonistic—a nihilistic—force. He didn't speak much about the censorship episode with his friends or colleagues, but this private knowledge unsettled him.

Nobody else seemed to share his anxiety. By all appearances, plans for major policy continued to advance rapidly. After the Johnston hearing, Timothy Wirth, a freshman Democratic senator from Colorado on the energy committee, began to plan a comprehensive package of climate change legislation—a New Deal for global warming. Wirth asked a legislative assistant, David Harwood, to consult with experts on the issue, beginning with Rafe Pomerance, in the hope of transforming national energy policy.

In March 1988, Wirth joined forty-one other senators, nearly half of them Republicans, to demand that Reagan pursue an international treaty modeled after the ozone agreement. Because the United States and the Soviet Union were the world's two largest contributors of carbon emissions, responsible for about one-third of the global total, they should lead the negotiations. Reagan agreed. In May, he signed a joint statement with Mikhail Gorbachev that included a pledge to cooperate on global warming.

But a pledge didn't reduce emissions. Hansen was learning to think more strategically—less like a scientist and more like a politician. Despite Wirth's efforts, there was as yet no serious national or international plan to limit fossil fuel consumption. Even Al Gore himself had, for the moment, withdrawn his claim to the issue. In 1987, at the age of thirty-nine, Gore had announced that he was running for president, in part to bring attention to global warming, but when the subject failed to thrill primary voters in New Hampshire, he stopped mentioning it. Instead he spoke about the Palestine Liberation Organization (he didn't think the United States should negotiate with it), school prayer (he supported it), and the federal funding of abortion for low-income women (he opposed it). In April he dropped out of the race.

As spring turned to summer, Anniek Hansen noticed a change in her husband. He grew pale and unusually thin, almost gaunt. When she would ask him about his day, Hansen would reply with some ambiguity and automatically turn the conversation to sports: the Yankees, his children's teams. Even by his own standards, he was tense, quiet, removed; in the hours he spent at home, he was not really there. She would begin a conversation and find out that he hadn't heard a word she said. She knew what he was thinking: he was running out of time. We were running out of time.

Then came the summer of 1988 and Jim Hansen wasn't the only one who could tell that time was running out.

Part III

You Will See Things That You Shall Believe

1988-1989

Rafe Pomerance (center) and Daniel Becker (far right) at Noordwijk in 1989

14. Nothing but Bonfires

Summer 1988

It was the hottest and driest summer in history. Everywhere you looked, something was bursting into flames. Two million acres in Alaska were incinerated, and dozens of major fires scored the West. Yellowstone National Park lost nearly a million acres. Smoke was visible from Chicago, sixteen hundred miles away.

In Nebraska, suffering its worst drought since the Dust Bowl, there were days when every weather station in the state registered temperatures above 100 degrees. The director of the Kansas Department of Health and Environment warned that the drought might be the dawning of a climatic change that within a half century could turn the state into a desert. “The dang heat,” said a farmer in Grinnell. “Farming has so many perils, but climate is ninety-nine percent of it." In parts of Wisconsin, where Governor Tommy Thompson banned fireworks and smoking cigarettes outdoors, the Fox and Wisconsin Rivers evaporated until they could not assimilate the sewage discharged into them. "At that point," said an official from the Department of Natural Resources, "we must just sit back and watch the fish die."

Harvard University, for the first time, closed because of heat. New York City's streets melted, its mosquito population quadrupled, and its murder rate reached a record high. "It's a chore just to walk," a former hostage negotiator told a reporter. "You want to be left alone." The twenty-eighth floor of Los Angeles's second-tallest building burst into flames; the cause, the fire department concluded, was spontaneous combustion. Ducks fled the continental United States in search of wetlands, many ending up in Alaska, swelling its pintail population to 1.5 million from 100,000. "How do you spell relief?" asked a spokesman for the Fish and Wildlife Service. "If you are a duck from America's parched prairies, this year you may spell it A-L-A-S-K-A."

Nineteen Miss Indiana contestants, outfitted with raincoats and umbrellas, sang "Come Rain or Come Shine," but it did not rain. The Reverend Jesse Jackson, who had outlasted Al Gore in the Democratic primary, stood in an Illinois cornfield and prayed for rain, but it did not rain. Cliff Doebel, the owner of a gardening store in Clyde, Ohio, paid $2,000 to import Leonard Crow Dog, a Sioux Indian medicine man from Rosebud, South Dakota. Crow Dog had performed 127 rain dances, all successful. "You will see things that you shall believe," he told the townspeople of Clyde.

"You will feel there is a chance for us all." After three days of dancing, it rained almost a quarter of an inch.

Texas farmers fed their cattle cacti. Stretches of the Mississippi River flowed at less than one-fifth of normal capacity. At Greenville, Mississippi, 1,700 barges beached; an additional 2,000 were marooned at St. Louis and Memphis. The on-field thermometer at Veterans Stadium in Philadelphia, where the Phillies were hosting the Chicago Cubs for a matinee, read 130 degrees. During a pitching change, every player, coach, and umpire, save the catcher and the entering reliever, Todd Frohwirth, fled into the locker rooms. (Frohwirth would earn the victory.) In the Cleveland suburb of Lakewood on June 21, yet another record smasher, a roofer working with 600-degree tar screamed, "Will this madness ever end?"

On June 22 in Washington, where it hit 100 degrees, Rafe Pomerance received a call from Jim Hansen, who was scheduled to testify the following morning at a Senate hearing called by Timothy Wirth.

"I hope we have good media coverage tomorrow," said Hansen.

This tickled Pomerance. He was the one who tended to worry about press; Hansen usually claimed indifference to such vulgar considerations. "Why's that?" he asked.

Hansen explained that he had just received the most recent global temperature data. Barely over halfway into the year, 1988 was setting records. Already it had nearly clinched the hottest year in history. Ahead of schedule, the signal was emerging from the noise.

"I'm going to make a pretty strong statement," said Hansen.

15.

Signal Weather

June 1988

The night before the hearing, Hansen flew to Washington to give himself enough time to prepare his oral testimony in his hotel room. But he couldn't focus—the ball game was on the radio. The slumping Yankees, who had fallen behind the Tigers for first place, were trying to avoid a sweep in Detroit, and the game was tight, a pitchers' duel between John Candelaria and Frank Tanana that went into extra innings. Hansen fell asleep without finishing his statement.

Hansen had told Pomerance that the biggest problem with the previous hearing, at least apart from the whole censorship affair, had been the month in which it was held: November. "This business of having global warming hearings in such cool weather," he said, "is never going to get attention." He wasn't joking. But they wouldn't have that concern again. On the morning of the Wirth hearing, Hansen awoke to bright sunlight, high humidity, choking heat. It was signal weather in Washington: the hottest June 23 in history.

Before heading to the Capitol, he attended a meeting at NASA headquarters. One of his early champions at the agency, Ichtiaque Rasool, was announcing the creation of a new carbon dioxide program. Hansen, sitting in a room with about thirty other scientists, continued to scribble his testimony under the table, barely listening. He paused, however, when he heard Rasool say that the program aimed to determine when a warming signal might emerge. "As you all know," said Rasool, "no respectable scientist would say that you already have a signal."

Hansen interrupted.

"I don't know if he's respectable or not," he said, "but I do know one scientist who is about to tell the U.S. Senate that the signal has emerged."

The other scientists looked up in surprise, but Rasool ignored Hansen and continued his presentation. Hansen returned to his testimony. He wrote: "The global warming is now large enough that we can ascribe with a high degree of confidence a cause-and-effect relationship to the greenhouse effect." He wrote: "1988 so far is so much warmer than 1987, that barring a remarkable and improbable cooling, 1988 will be the warmest year on record." He wrote: "The greenhouse effect has been detected, and it is changing our climate *now*."

By 2:10 p.m., when the session began, it was 98 degrees,

and not much cooler in Room 366 of the Dirksen Senate Office Building, which was being irradiated by two banks of television-camera lights. The reporters had been alerted by Timothy Wirth's office that the plainspoken NASA scientist was going to make a "major statement." After the congressional staff saw the cameras, even those senators who hadn't planned to attend appeared at the dais, hastily reviewing the remarks their aides had drafted for them.

Half an hour before the hearing, Wirth pulled Hansen aside. Wirth wanted to change the order of speakers, placing Hansen first. The senator wanted to make sure that his statement got sufficient attention. Hansen agreed.

"We have only one planet," Senator Bennett Johnston intoned. "If we screw it up, we have no place to go." Senator Max Baucus, a Democrat from Montana, called for the United Nations Environment Programme to begin preparing a global remedy to the carbon dioxide problem. Senator Dale Bumpers, a Democrat from Arkansas, previewed Hansen's testimony, saying that it "ought to be cause for headlines in every newspaper in America tomorrow morning." Press coverage, Bumpers emphasized, was a necessary precursor to policy. "Nobody wants to take on any of the industries that produce the things that we throw up into the atmosphere," he said. "But what you have are all these competing interests pitted against our very survival."

Wirth asked those standing in the gallery to claim the few remaining seats available. "There is no point in standing up through this on a hot day," he said, glad for the opportunity to advertise the weather. Then he introduced the star witness.

Hansen, wiping his brow, spoke without affect, his eyes rarely rising from his notes. The warming trend could be detected "with ninety-nine percent confidence," he said. He encouraged the senators to do what was possible to curb the warming immediately. But he saved his strongest words for after the hearing, when he was smothered by reporters. "It is time to stop waffling so much," he said, "and say that the evidence is pretty strong that the greenhouse effect is here."

The press followed Bumpers's advice. Hansen's testimony led the national news; across the top of its front page, *The New York Times*' headline read "Global Warming Has Begun." The images of Hansen in the Dirksen hearing room were ubiquitous, in halftone and high gloss and pixel. The television broadcasts showed Hansen hunched over his testimony, the shoulders of his beige suit bunched up, reading in his deliberate monotone; print publications used a *Mr. Smith Goes to Washington* black-and-white portrait shot by a cameraman lying below the witness table, in which Hansen, jaw clenched, teeth gritted, tie tightly knotted, stares down some questioning senator, delivering his grim verdict. For the first time the crisis had a face and, with it, an emotion—a fever of pent-up frustration, outrage, and moral conviction.

But Hansen had no time to dwell on any of this. When he got home to New York, Anniek told him that she had breast cancer. She had found out two weeks earlier but hadn't wanted to upset him before the hearing. In the following days, while the entire world tried to learn about James Hansen, he tried to learn about Anniek's illness. He took her for a second opinion. After he absorbed the initial shock and made a truce with the fear—his grandmother had died from the disease, orphaning his mother at the age of twelve—he dedicated himself to Anniek's treatment with all the rigor he brought to his research. He stayed home from the Goddard Institute, helping around the house and preparing their children for school. He studied mammograms and ultrasounds. As they weighed treatment options and analyzed medical data, Anniek noticed him begin to change. The frustration of the last year began to fall away. It yielded, in those doctors' offices, to a steady coolness, an obsession for detail, a dogged optimism. He began to look like himself again.

16.

Woodstock for Climate Change

June 1988–April 1989

In the immediate flush of optimism after the Wirth hearing (henceforth known as the Hansen hearing), Rafe Pomerance called his allies on Capitol Hill—the young staffers who advised politicians, organized hearings, wrote legislation. We need a number, he told them, a specific target in order to move the issue, to turn all this publicity into policy. The Montreal Protocol had called for a 50 percent reduction by 1998 in CFC emissions. What was the right target for carbon emissions? It wasn't enough to exhort the world's nations to do better. That kind of lofty talk might sound noble, but it didn't change investments or laws. They needed a hard goal—something ambitious but reasonable. And they needed it soon: just four days after Hansen's star turn, politicians from forty-six nations and more than three hundred scientists would convene in Toronto at the World Conference on the Changing Atmosphere, an event *The New York Times*' Philip Shabecoff called "Woodstock for climate change."

Pomerance hastily arranged a meeting with David Harwood, the architect of Wirth's climate legislation; Roger Dower in the Congressional Budget Office, who at Wirth's request was calculating the plausibility of a national carbon tax; and Irving Mintzer, a colleague at the World Resources Institute who had a deep knowledge of energy economics. Wirth was scheduled to give the keynote address in Toronto—Harwood would write it—and the senator could propose a number then. But which number?

Pomerance had one in mind: a 20 percent reduction in carbon emissions by 2000.

Ambitious, said Harwood. In all his work planning climate policy, he had seen no assurance that such a steep drop in emissions was possible. Then again, 2000 was more than a decade off, so it gave them some rope.

What really mattered wasn't the number itself, said Dower, but that they picked one. He agreed that a hard target was the only way to push the issue forward. Though his job at the Congressional Budget Office required him to come up with precise estimates of speculative, complex policy, there wasn't time for yet another academic study to arrive at the exact right number. Pomerance's unscientific suggestion sounded fine to him.

Mintzer pointed out that it was consistent with the academic literature on energy efficiency: most energy systems could be improved by roughly 20 percent if you adopted best practices. Of course, with any emissions target, you had to take into account the fact that the developing world, with its booming population growth, would inevitably ingest much larger quantities of fossil fuels by 2000. But those gains could be offset by a wider propagation of the renewable technologies already at hand—solar, wind, geothermal. It was not a rigorous scientific analysis, Mintzer granted, but 20 percent sounded plausible. We wouldn't need to solve cold fusion or ask Congress to repeal the law of gravity. We could manage it with the knowledge and technology we already had.

Besides, said Pomerance, *20 by 2000* sounds good.

In Toronto a few days later, Pomerance talked up his idea with everyone he met—environmental ministers, scientists, journalists. Nobody thought it sounded crazy. He took that as an encouraging sign. Other delegates soon proposed the number to him independently, as if they had come up with it themselves. That was an even better sign.

Wirth, in his keynote on June 27, called for the world to reduce emissions by 20 percent by 2000, with an eventual reduction by 50 percent. It was the first specific benchmark to be proposed at a major international meeting. Other speakers likened the ramifications of climate change to a global nuclear war, but it was the emissions target that sold in Washington, London, Berlin, and Moscow. The conference's final statement, signed by all four hundred scientists and politicians in attendance, repeated the demand with a slight variation: a 20 percent reduction in carbon emissions by 2005. Just like that, Pomerance's hunch became global diplomatic policy.

Hansen, emerging from Anniek's successful cancer surgery, took it upon himself to start a one-man global warming public information campaign. He gave news conferences and was quoted in seemingly every article about the issue; he even appeared on television with homemade props. Like an entrant in an elementary school science fair, he made "loaded dice" out of sections of cardboard, Scotch tape, and colored paper to illustrate the increased likelihood of hotter weather in a warmer climate. Public awareness of the greenhouse effect reached an all-time high of 68 percent. Global warming caused approximately one-third of Americans to worry "a great deal."

At the end of the sulfurous summer, several months after Gore suspended his candidacy, the climate crisis at last became a major issue in the presidential campaign. When Michael Dukakis proposed tax incentives to encourage domestic oil production and boasted that coal could satisfy the nation's energy needs for the next three centuries, George H. W. Bush took advantage. "I am an environmentalist," he declaimed on the shore of Lake Erie, the first stop on a five-state environmental tour that would take him to Boston Harbor, Dukakis's home turf. "Those who think we are powerless to do anything about the greenhouse effect," he said, "are forgetting about the White House effect." His running mate emphasized the ticket's commitment to solving global warming at the vice presidential debate. "The greenhouse effect is an important environmental issue," said Dan Quayle. "We need to get on with it. And in a George Bush administration, you can bet that we will." A week before the election, barnstorming in Alabama, Bush interrupted a speech on space exploration with a mournful reflection on global warming and the fate of Earth, summoning his inner Aldo Leopold: "I recall the old fisherman's prayer: 'The sea is so large, Lord, and my boat is so small . . .' We face the prospect of being trapped on a boat we have irreparably damaged, not by the cataclysm of war but by the slow neglect of a vessel we believed to be impervious to our abuse."

Two weeks after the election, Bush was visited in the White House by former presidents Jimmy Carter and Gerald Ford. They presented him with *American Agenda*, a year-long, bipartisan report on the challenges facing the country. It recommended making climate change a major national priority and doubling the EPA's research budget. Americans, the former presidents said, "should no longer see environmental issues simply as a luxury."

A lawyer for the Senate energy committee told an industry journal that lawmakers were "frightened" by Bush's commitment to the issue and predicted that Congress would pass significant legislation once he took office. "A lot of people on the Hill see the greenhouse effect as the issue of the 1990s," a gas lobbyist told *Oil & Gas Journal*. The coal industry, which had the most to lose from restrictions on carbon emissions, had already moved beyond denial to acceptance. The National Coal Association acknowledged that the greenhouse effect was no longer "an emerging issue. It is here already, and we'll be hearing more and more about it."

By the end of the year, thirty-two climate bills had been introduced in Congress, led by two gigantic omnibuses: the

Global Warming Prevention Act, proposed in the House by Claudine Schneider, a Republican from Rhode Island; and Wirth's National Energy Policy Act of 1988, cosponsored by thirteen Democrats and five Republicans. Both called for the establishment of an "International Global Agreement on the Atmosphere" by 1992; Wirth's called for a 20 percent reduction of carbon dioxide in the atmosphere by 2005, a reduction of national energy use by at least 2 percent every year until then, and $1.5 billion in funding for worldwide birth control. At a meeting of oil executives shortly after the election, Representative Dick Cheney, a Republican from Wyoming, warned that a gasoline tax would "be very difficult to fend off" after Bush took office.

The other great powers refused to wait that long. The German parliament created a special commission on climate change, which called the Toronto goal inadequate and concluded that action had to be taken immediately, "irrespective of any need for further research"; it recommended a 30 percent reduction of carbon emissions. Sweden's Parliament announced a national strategy to stabilize emissions at the 1988 level and later imposed a carbon tax. Margaret Thatcher, who had studied chemistry at Oxford, warned in a speech to the Royal Society that global warming could "greatly exceed the capacity of our natural habitat to cope" and that "the health of the economy and the health of our environment are totally dependent upon each other."

It was at this moment—when the environmental movement was, in the words of one energy lobbyist, "on a tear"—that the United Nations unanimously endorsed the establishment, by UNEP and the WMO, of the Intergovernmental Panel on

Climate Change, to be composed of scientists and policy-makers who would conduct scientific assessments and develop a global climate policy.

During the transition period, Bush's administration invited the IPCC's Response Strategies Working Group, the body responsible for planning a climate treaty, to hold one of its first sessions at the U.S. State Department. Bush had promised to combat the greenhouse effect with the White House effect. The self-proclaimed environmentalist would soon be seated in the Oval Office. It was time.

17.

Fragmented World

Fall 1988

The oilmen started calling Terry Yosie.

"What is this global warming business all about, Terry?"

"What's the story with this guy Hansen, Terry?"

"What are we doing about this, Terry?"

Yosie was ready for them. Though he had only recently been hired by the American Petroleum Institute to run its health and environment division, he was on easy terms with the CEOs of most major oil and gas companies, having worked at the EPA for the past seven years. Ozone was his expertise, though he had helped negotiate the joint statement on global warming signed by Reagan and Gorbachev in Moscow. After the Hansen hearing, it had become clear to

Yosie that nobody at API knew much about the subject. His boss, executive vice president William O'Keefe, who had worked there since 1979 and had run the health and environment division for the previous five years, had never even heard of it. At least not until James Hansen came along.

In all the wildness and havoc that followed the hearing, API, unprepared to provide comment, had offered the same boilerplate it used for any new environmental concern, whether benzene or ozone or smog: it cited a "divergence of expert opinion" and warned that "premature actions taken to address this issue could be disruptive and a potential waste of society's resources." *More research was needed.* That was Yosie's assignment—to do the research.

Some of API's members, he came to learn, had begun, however tentatively, to do their own research. British Petroleum, which had spent $11 billion on rigs, roads, and pipelines that rested on Alaskan permafrost, wanted to figure out what would happen if the frost wasn't perma: What if icebergs calved and gave oil tankers the *Titanic* treatment? Mobil, too, had advanced beyond a posture of skepticism. In November 1988, its president, Richard Tucker, speaking at a national conference for the American Institute of Chemical Engineers, warned that action to address the greenhouse effect might require "a dramatic reduction in our dependence on fossil fuels." And in spring 1988, Royal Dutch Shell had released a rigorous internal briefing report on the greenhouse effect, initiated more than two years earlier, laying out the science, the social and economic effects, and the challenges of achieving an effective solution. The authors concluded that "the energy industry needs to consider how to play its part," but argued that the main burden would have to be assumed by governments, as "the only effective way to tackle the problem is through international cooperation."

Shell had questioned the prospects of international cooperation since at least 1982, however, when it hired the futurist Peter Schwartz to run its long-term strategy division. Schwartz had previously worked at the Stanford Research Institute, where, in 1977, he oversaw the grim sixty-page report, commissioned by the Energy Research and Development Administration, on the sociopolitical impacts of global warming. At Shell, Schwartz had developed two divergent models of the future. In "The Next Wave," climate change prompted deep investments in renewable energy; in "Fragmented World," the major nations developed differing responses to climate change, leading to geopolitical turmoil and, in aggregate, an increasing reliance on fossil fuels.

Exxon used different terminology, but by the summer of 1988 it had begun to formulate its own strategy of bringing about a fragmented world. After Hansen's hearing, Duane LeVine, Exxon's manager of science and strategy development, requested a new corporate position on global warming—not from his scientists, but from a strategist, Joseph Carlson, a public relations officer. LeVine wanted Carlson to answer one question in particular: "What do you think is the direct impact on Exxon's business?" Carlson's first draft, written in August without consulting his corporate superiors, accepted the consensus view of the science. But he proposed that Exxon should resist any efforts at "overstatement and sensationalization," which "could lead to noneconomic development of nonfossil fuel resources." It would be useful, he argued, to "emphasize the uncertainty in scientific conclusions regarding the potential enhanced Greenhouse effect."

Yet Terry Yosie found that most of the industry, particularly the nonscientists, were altogether ignorant of the issue and its potential economic significance. With the encouragement of API's president, Charles DiBona, Yosie organized a workshop to be hosted in a lecture hall within the institution's L Street headquarters. Nearly a hundred people attended—men in chalked haberdashery paid to scan the horizon for opportunity and risk; senior lobbyists and lawyers from Shell, Arco, Chevron, and Mobil; and a retinue of senior API executives. Yosie invited guest speakers who he thought could speak credibly about the science and policy ramifications: Brian Flannery, Exxon's in-house authority on climate modeling, who explained that climate change was real and that fossil fuel combustion was the cause; a political scientist, Aaron Wildavsky, who warned that regulations on energy production could hurt profits; Richard Morgenstern, the director of the EPA's Office of Policy Analysis, who emphasized that the costs of controlling climate change were still fairly modest and worth pursuing; and Irving Mintzer of the World Resources Institute.

Before the meeting, Mintzer consulted with his colleague Rafe Pomerance to sharpen his argument. Mintzer, like Morgenstern, told the audience at API that immediate, prudent actions—like investing in efficiency and renewable energy—would be economically beneficial. The economic risks of inaction, on the other hand, were extreme: a 2-degree warming could stagger the global economy in ways not yet imagined.

The longer the industry waited to act, the worse it would go for them. The audience took this in stride.

After the presentation DiBona asked Yosie to give a briefing to API's executive committee, composed of about fifteen CEOs of the world's largest oil and gas companies. Yosie prepared a speech and a slideshow. There was no doubt that oil and gas combustion was making the world hotter, he explained. Yes, he granted, uncertainties remained, such as the timing of the changes. But the trend, as Exxon's Flannery had explained, was not in doubt. He showed the world's great oil barons the Keeling curve; the data quantifying the contribution of fossil fuels to global warming; and a chart of worldwide emissions, separated out by region. He also included a statistic rarely mentioned in accounts of the science. A warming world, he explained, would stimulate greater energy use, mainly due to higher demand for air-conditioning and refrigeration. By 2055, he told the executives, climate change would increase national energy consumption by 4 to 6 percent.

What should be done? *What was the direct impact on business?* As Yosie saw it, three strategic options lay available to them. The first was a binding global treaty, requiring strong intervention from governments. The second amounted to a shrug: Yosie quoted John Maynard Keynes's observation that "in the long run we'll all be dead," so why worry? The final position, which Yosie endorsed, echoed William Nierenberg's message five years earlier, after the publication of *Changing Climate*: proceed with caution, not panic, making sure that regulatory policy was applied gradually, in order to avoid any economic shocks. The best way to do this, argued Yosie, was to make the industry "an active participant in the scientific and policy debate." It should highlight uncertainties in the science, question the effectiveness of any new regulations, urge international cooperation, and accept only those measures "consistent with broader economic goals," which is to say, actions that didn't hurt profits.

The oilmen agreed. They set aside money for policy analysis—about $100,000, a fraction of the environmental division's approximately $30 million budget. API, by comparison, spent millions of dollars a year to study the health effects of benzene. But even a few thousand dollars would help. It was enough to get a press campaign going. It was enough to show the world that the industry cared.

At the end of 1988, API's president began to audition its policy arguments in the trades. At a briefing in December, described as a year-end "pep talk" that reviewed the anticipated energy policies of the Bush administration, DiBona raised the specter of global warming legislation. "Many people are already using the 'greenhouse' fever to push agendas built around extreme environmental and conservation ideas," he said. "Unless cooler, less biased heads prevail, the nation could scare itself into a costly, nearly impossible set of environmental goals, with tremendous burdens on U.S. industry and society." DiBona acknowledged the scientific consensus that carbon dioxide was rising and would warm the planet. But scientists couldn't say with certainty how quickly the warming would occur. It was important, he emphasized, that whatever happened next, the industry had to stand together. The industry journalists dutifully marked down his words.

18.

The Great Includer and the Old Engineer

Spring 1989

The IPCC's Response Strategies Working Group convened at the State Department ten days after Bush's inauguration to begin the process of negotiating a global treaty. James Baker III chose the occasion to make his first speech as secretary of state. He had received a memo from Frederick M. Bernthal, a former nuclear regulatory commissioner and chemistry professor who was an assistant secretary of state for international environmental affairs and had been named the chairman of the IPCC working group. In frank, brittle prose, Bernthal argued that it was prudent to begin a reduction in greenhouse gas emissions; the costs of inaction were simply too high. Baker adopted some of Bernthal's language in his speech. "We can probably not afford to wait until all of the uncertainties about global climate change have been resolved before we act," said Baker. "Time will not make the problem go away."

After the speech, Baker received a visit from John Sununu, Bush's chief of staff.

"Leave the science to the scientists," Sununu told Baker. "Stay clear of this greenhouse effect nonsense. You don't know what you're talking about."

Baker, who had served as Reagan's chief of staff and Treasury secretary, didn't speak about the subject again. He later told the White House that he had recused himself from consulting on energy policy, on account of his previous career as a Houston oil and gas lawyer.

Bush had chosen Sununu for his political instincts—he was credited with delivering the New Hampshire primary after Bush had come in third in Iowa and securing Bush's nomination. But despite his reputation as a political wolf, and his three terms as New Hampshire governor, Sununu still thought of himself as an "old engineer," as he was fond of putting it, having earned a Ph.D. in mechanical engineering from MIT decades earlier. He took great pleasure in defying others' lazy characterizations of himself; he was an enthusiastic contrarian and a charming bully. His father was a Lebanese exporter from Boston, his mother a Salvadoran of Greek ancestry, and he was born in Havana. In New Hampshire he had come, in the epithets of national political columnists, to embody Yankee conservatism: pragmatic, business-friendly, technocratic, "no-nonsense." He had fought angrily against local environmentalists to open a nuclear power plant, but he had also signed the nation's first acid-rain legislation and lobbied Reagan in person for a 50 percent reduction in sulfur dioxide pollution, the target sought by the Audubon Society. He increased spending on mental health care and public land preservation. Still he was considered more conservative than Reagan, a budget hawk who had turned a $44 million state deficit into a surplus without raising taxes. Sununu openly insulted Republican politicians and the president of the U.S. Chamber of Commerce when they drifted, however subtly, from his anti-tax doctrinairism. Once in the White House, however, he would help negotiate a tax increase and secure the Supreme Court nomination of David Souter.

As an old engineer, Sununu lacked the reflexive deference that so many of his political generation reserved for the vaunted class of elite government scientists. Though he had served as an adviser at MIT's graduate program in technology and public policy, he harbored skepticism toward scientists who mingled the two professionally. During his years in government he had nursed this skepticism into a full-fledged theory of twentieth-century geopolitics. Since World War II, he believed, conspiratorial forces had used the imprimatur of scientific knowledge to advance a socialistic, "anti-growth" doctrine. He reserved particular disdain for Paul Ehrlich's *The Population Bomb*, which prophesied that hundreds of millions of people would starve to death if the world failed to curb population growth; the Club of Rome, an organization of European scientists, heads of state, and economists, which warned that the world would run out of natural resources; and, as recently as the mid-seventies, the hypothesis advanced by some of the nation's leading authorities on global climate—among them Carl Sagan, Stephen Schneider, and Ichtiaque Rasool—that a new ice age was dawning, thanks to the proliferation of synthetic aerosols. All were theories of questionable scientific merit, portending vast, authoritarian remedies to halt economic progress, and all had been debunked.

Sununu had suspected that the greenhouse effect belonged to this nefarious cabal since 1975, when Margaret Mead spoke out on the subject. "Never before have the governing bodies of the world been faced with decisions so far-reaching," she wrote. "It is inevitable that there will be a clash between those concerned with immediate problems and those who concern themselves with long-term consequences." When Mead talked about "far-reaching" decisions and "long-term consequences," Sununu heard the marching of jackboots.

On April 14, 1989, a bipartisan group of twenty-four senators, headed by majority leader George Mitchell, requested that Bush cut carbon emissions in the United States even before the IPCC's working group made its recommendation. "We cannot afford the long lead times associated with a comprehensive global agreement," the senators wrote. Sununu learned from Richard Darman, the director of the Office of Management and Budget, a close ally, that Al Gore was going to call a hearing to shame Bush into taking immediate action. James Hansen would again serve as lead witness. Darman had the testimony on his desk and described it. Sununu was appalled: Hansen's warnings seemed extreme to him, especially since they were based on scientific arguments that Sununu considered, as he put it, "technical garbage."

While Sununu and Darman reviewed Hansen's statements, the recently sworn-in EPA administrator, William K. Reilly, brought a new proposal to the White House. Reilly—tall, patrician, direct and pragmatic of manner, crisply attired, bassoon-voiced, a former army captain and intelligence officer with degrees from Yale, Harvard, and Columbia—was a professional environmentalist, having served as a staffer on Nixon's Council on Environmental Quality and later as president of the Conservation Foundation and the World Wildlife Federation. For his dogged pursuit of political consensus, he had been nicknamed "the Great Includer"; his nomination to the EPA was praised by the National Coal Association and the National Resources Defense Council, the Chemical Manufacturers Organization and the Sierra Club. At his swearing-in, Bush boasted that the selection of Reilly should make it "plain to everyone in this room and around the country that among my first items on my personal agenda is the protection of America's environment." One of the first items on Reilly's own agenda was global warming. The next assembly of the IPCC's working group was scheduled to take place in Geneva the following month, in May. It was the perfect occasion, Reilly argued in a meeting at the White House, to prove that the administration was serious about the greenhouse effect. Bush should demand a global treaty to reduce carbon emissions.

Sununu disagreed. It would be foolish, he said, to let the nation stumble into a binding agreement on questionable scientific merits, especially one that would compel some unknown quantity of economic pain. They went back and forth. Reilly didn't want to cede leadership on the issue to the European powers; just a few months later, after all, Reilly would travel to the Netherlands to attend the first high-level diplomatic meeting on climate change. Statements of caution would make the "environmental president" look like a hypocrite and undermine the United States' leverage in a negotiation. But Sununu wouldn't budge. He ordered the U.S. delegates not to make any commitment in Geneva. Soon after, someone leaked the exchange to the press.

Their dispute was reported by *The Washington Post* and the AP ahead of the Geneva meeting, characterizing Sununu as single-handedly thwarting the will of various "top Bush administration officials," led by a "frustrated" Reilly. Sununu, furious, was certain that Reilly was the source. Spilling to the press, especially about internal disputes, was intolerable. The articles made the administration look as if it didn't know what it was doing. It drove Sununu crazy.

Jim Baker's deputy pulled Reilly aside. He said he had a message from Baker. "In the long run," the deputy warned Reilly, "you never beat the White House." Sununu, Baker predicted, wouldn't forget this.

19.

Natural Processes

May 1989

In the first week of May, when Hansen received his proposed congressional testimony back from the White House, it was disfigured by deletions and, more incredible, substantial additions. Gore had called the hearing to increase pressure on Bush, but Hansen had agreed to testify for a different reason: he worried that one of the major points he had tried to make at the 1988 hearing had been misunderstood. His research had found that global warming would not only cause more heat waves and droughts like those of the previous summer; it would also lead to more extreme rain events. This was a crucial disclaimer—he didn't want the public to assume, the next time there was a mild summer, that global warming wasn't real.

But the revised text was a mess. For a couple of days, Hansen played along, accepting the more innocuous edits, though he refused to accept some of the howlers proposed by the Office of Management and Budget. With the hearing only two days away, he gave up. He told NASA's congressional liaison to stop fighting. This time, he wouldn't go through the charade of testifying as "Atmospheric Scientist, New York, N.Y." He told the liaison to let the White House have its way.

But Hansen would have his way too. As soon as he hung up, he drafted a letter to Gore. He explained that OMB wanted him to demote his own scientific findings to "estimates" from models that were unreliable and "evolving." His anonymous censor wanted him to say that the causes of global warming were "scientifically unknown" and might be attributable to "natural processes," caveats that would not only render his testimony meaningless but make him sound like a moron. The most bizarre addition, however, was a statement of a different kind. He was asked to demand that Congress consider only climate legislation that would immediately benefit the economy, "independent of concerns about an increasing greenhouse effect"—a sentence no scientist would ever utter, unless perhaps he was employed by the American Petroleum Institute. Hansen faxed his letter to Gore and left the office.

Upon arriving home, he learned from Anniek that Gore had called. When Hansen called back, Gore told him that he planned to tell a couple of reporters what had happened. Hansen said that was fine with him.

Philip Shabecoff of *The New York Times* called the next

morning. "I should be allowed to say what is my scientific position," Hansen told him. "I can understand changing policy, but not science."

On Monday, May 8, the morning of the hearing, Hansen left early for his flight to Washington and did not see the newspaper until he arrived at Dirksen. Gore showed it to him. The front-page headline read, "Scientist Says Budget Office Altered His Testimony." Gore prepared him to brace for another public bonfire. They agreed that Hansen would give his testimony as planned, after which Gore would ask about the rewritten passages.

Gore stopped at the door. "We better go separately," he said. "Otherwise they'll be able to get both of us with one hand grenade."

The official title of the hearing was "Climate Surprises," but the only surprise the press cared about was the White House's interference with Hansen's testimony. In the overstuffed hearing room, the cameras fixed on the reluctant celebrity scientist. Hansen held his statement in one hand and a single Christmas-tree bulb in the other—a prop to help explain, however shakily, that the warming already created by fossil fuel combustion was equivalent to placing a Christmas light over every square meter of Earth's surface. After he read his sanitized testimony, Gore pounced. He was puzzled by inconsistencies in Hansen's presentation, he said, in a tone thick with mock confusion. "Why do you directly contradict yourself?"

Hansen explained that he had not written those contradictory statements. He explained that he certainly did not agree, for instance, with the claim that his scientific findings were not reliable. He did not quarrel with the White House's practice of reviewing policy statements by government employees. But, he added, in the same flat tone he used to explain the phenomenon of La Niña, "my only objection is being forced to alter the science."

Gore was beside himself. "The Bush administration is acting as if it is scared of the truth," he said. "If they forced you to change a scientific conclusion, it is a form of science fraud." He worked himself into a righteous fervor. "You know, in the Soviet Union they used to have a tradition of ordering scientists to change their studies to conform with the ideology then acceptable to the state. And scientists in the rest of the world found that laughable as well as tragic."

Another government scientist testifying at the hearing, Jerry Mahlman from NOAA, acknowledged that the White House had tried to change his conclusions too. Mahlman had managed to deflect the worst of it, however—"objectionable and also unscientific" recommendations, he said, that would have been "severely embarrassing to me in the face of my scientific colleagues."

"The time has come," said Senator Timothy Wirth, for "greater confrontation" with the White House, "to push them and maybe embarrass them into action if, in fact, they cannot be led to action by a more reasonable approach."

Gore called it "an outrage of the first order of magnitude." The 1988 hearing had created a hero out of Jim Hansen. Now Gore had a real villain, one far more treacherous than Fred Koomanoff—a nameless censor in the White House, hiding behind OMB letterhead.

After recess, the stroboscopic lights followed Hansen and

Gore into the marbled hallway. Hansen insisted that he wanted to focus on the science. Gore was happy to focus on the politics. "I think they're scared of the truth," he said. "They're scared that Hansen and the other scientists are right and that some dramatic policy changes are going to be needed, and they don't want to face up to it."

At the White House press briefing later that morning, Press Secretary Marlin Fitzwater conceded that Hansen's statement had been altered. He blamed an official "five levels down from the top" and promised that there would be no retaliation against Hansen, who, he added, was "an outstanding and distinguished scientist" and was "doing a great job."

The episode did more to publicize the need for climate policy than any testimony Hansen could have delivered. It was "an outrageous assault" (*Los Angeles Times*) that marked the beginning of "a cold war on global warming" (*Chicago Tribune*), sending "the signal that Washington wants to go slow on addressing the greenhouse problem" (*The New York Times*). An administration that valued political posture above political policy was caught, once again, in a defensive crouch.

The day after the hearing, Gore received an unannounced visit from Richard Darman. He came alone, without aides. He said he wanted to apologize to Gore in person. He was sorry and he wanted Gore to know it; OMB would not try to censor anyone again. Gore, stunned, thanked Darman. It was wildly out of character: Dick Darman was notoriously conniving, arrogant; even fellow Republicans complained to reporters that he was "abrasive." He graduated Harvard a year before Gore entered, served as James Baker's deputy at the Treasury Department, and had been one of Reagan's master strategists. He was an insider's insider, a professional cynic, a hardened member of Capitol Hill's managerial elite. Yet Gore suspected his apology was sincere. The effusiveness, the mortified tone, and the fact that he had come unaccompanied, as if in secret, left Gore with the impression that the idea to censor Hansen didn't come from someone five levels down from the top, or even from Darman. It had come from someone above Darman.

20.

The White House Effect

Spring-Fall 1989

Dick Darman went to see Sununu. He didn't like being accused of censoring scientists. They needed to issue some kind of response. Sununu called Reilly to ask if he had any ideas. We could begin, said Reilly, by recommitting to a global climate treaty, as he had first suggested. The United States was the only Western nation on record as opposing negotiations.

On the evening of Thursday, May 11, three days after the disastrous hearing, and two days after it was reported that Margaret Thatcher's government had called on world leaders to organize a global warming convention as soon as possible, Sununu sent a telegram to U.S. negotiators in Geneva, where the IPCC meeting was already under way. The telegram reversed his previous directive to avoid making any commitments. Instead, he said, the Americans should work "to develop full international consensus on necessary steps to prepare for a formal treaty-negotiating process. The scope and importance of this issue are so great that it is essential for the U.S. to exercise leadership." He further proposed an international workshop on global warming, to be hosted by the White House, that would aim to improve the accuracy of the science and calculate the economic costs of emissions reductions. Sununu signed the telegram himself. It wasn't enough for Al Gore: "Once again, the president has been dragged slowly and reluctantly toward the correct position. Although this is progress, it is still not nearly good enough." Nor did it satisfy Rafe Pomerance, who told reporters that the belated effort to save face was a "waffle" that fell short of real action. But the general public response was one of praise, and relief.

Still Sununu seethed at any mention of the subject. He had taken it upon himself to make a formal study of the greenhouse effect; he would have a rudimentary, one-dimensional general circulation model installed on his personal desktop computer. He decided that Hansen's models were horribly imprecise, "technical poppycock" that failed adequately to account for the ocean's capacity to mitigate warming. He complained about them to Darman and D. Allan Bromley, a nuclear physicist from Yale whom, at Sununu's recommendation, Bush had named science adviser. Hansen's models, Sununu groused, didn't begin to justify such wild-eyed pronouncements as "the greenhouse effect is here" or that the 1988 heat waves could be attributed to global warming. God forbid they be used as the basis for national economic policy. Darman and Bromley nodded along.

When a junior staffer in the Energy Department, in a White House meeting with Sununu and Reilly, mentioned in passing an initiative to reduce fossil fuel use, Sununu interrupted her.

"Why in the world would you need to reduce fossil fuel use?"

"Because of climate change," the young woman replied, uncertain.

Sununu went incandescent. "I don't want anyone in this administration without a scientific background using 'climate change' or 'global warming' ever again," he said. "If you don't have a technical basis for policy, don't run around making decisions on the basis of newspaper headlines."

After the meeting, Reilly caught up to the staffer in the hallway. She was shaken.

"Don't take it personally," Reilly told her. "Sununu might have been looking at you. But that was directed at me."

Relations between Sununu and Reilly became openly adversarial. Reilly had worked well with real estate developers and executives from the chemical and energy industries. But he had never encountered anyone like Sununu. Reilly's conservative bona fides meant nothing to Sununu, who considered him a creature of the environmental lobby: a lawyer and an urban planner by training who was trying to impress his pals at the EPA without having a basic grasp of the science. Most unforgivable of all was Reilly's propensity to gossip to the press any time internal decisions went badly for him. Whenever Reilly submitted to the White House candidates for openings at the EPA, Sununu vetoed them. He didn't trust Reilly to negotiate on behalf of the administration, so when it came time for the major conference in the Netherlands, where it was expected that the world's environmental ministers would endorse the IPCC process, Sununu decided to send Allan Bromley to accompany Reilly.

Reilly, ever conciliatory, didn't entirely blame Sununu for Bush's indecision on a climate treaty. The president had never taken a particularly vigorous interest in global warming. He had not discussed the issue at depth with scientists. (When scientists offered to brief the White House, they reported to Sununu.) Bush had brought up global warming on the campaign trail only after hunting through a briefing booklet for a new issue that might get him some positive press. When Reilly tried in person to persuade him to take action, Bush deferred to Sununu and Baker. Why don't the three of you work it out, the president said. Let me know when you decide.

But by the time Reilly landed in the Netherlands, he suspected that it was already too late.

21.

Skunks at the Garden Party

November 1989

Rafe Pomerance awoke at sunlight and stole out of his hotel, making for the flagpoles. It was nearly freezing—November 6 on the coast of the North Sea in the Dutch resort town of Noordwijk—but the wind had yet to gust and the photographer was waiting. More than sixty flags lined the strand between the hotel and the beach, one for each nation in attendance at the first major diplomatic summit on global warming. The environmental ministers would review the progress made by the IPCC and decide whether to endorse a framework for a global treaty. There was a general sense among the delegates that they would agree to the target proposed by the host, the Dutch minister—freezing greenhouse gas emissions at 1990 levels by 2000—which, after all, was more modest than the Toronto number. If the meeting was a success, it might encourage the IPCC to accelerate its negotiations and finalize a treaty more quickly. At the very least, the ministers planned to approve a binding target of emissions reductions. The energy in the seaside hotel was high, nearly giddy. After more than a decade of fruitless international meetings, they would finally sign an agreement that meant something.

Pomerance had not been among the four hundred delegates invited to Noordwijk. But together with three young activists—Daniel Becker of the Sierra Club, Alden Meyer of the Union of Concerned Scientists, and Stewart Boyle from Friends of the Earth—he had formed his own impromptu delegation. Their constituency, they liked to say, was the climate itself. Their mission was to pressure the delegates to include in the final conference statement, which would be used as the basis for a global treaty, the target proposed in Toronto: a 20 percent reduction of greenhouse gas combustion by 2005. It was the only measure that mattered, the amount of emissions reductions, and the Toronto number was the strongest target yet to have been widely embraced. With the ministers' endorsement, it would be a step closer to becoming global law.

The activists booked their own travel and doubled up in rooms at a beat-up motel down the beach. They managed to secure all-access credentials from the Dutch environmental ministry's press secretary. He was inclined to be sympathetic because it had been rumored that Allan Bromley, who was tagging along with William Reilly, might try to persuade the delegates from Japan and the Soviet Union to join him in resisting the idea of a binding agreement. There had been concern in recent weeks that something like this might happen; "Sununu is winning," Senator Wirth had told *The Washington Post*. On October 18, John Chafee and four other Republican senators (Rudy Boschwitz of Minnesota, Slade Gorton of Washington, James Jeffords of Vermont, and Robert Packwood of Oregon) wrote a stern, at times patronizing letter to Bush about the Noordwijk meeting. They urged him to direct his negotiators to propose a "forceful and specific agenda" on global warming. "Unless you provide personal leadership on this issue," they wrote, "the United States will continue to send mixed signals to the world community and will put forth proposals that will be subject to criticism at home and abroad." A successful negotiation, they stipulated, had to include commitments to freeze U.S. carbon dioxide emissions at current levels, establish specific targets for reductions, and assist developing countries to use renewable sources of energy. If the government failed to enact "an aggressive domestic policy on carbon dioxide emissions," it could not expect other nations to act accordingly. The Republican senators called their proposal "the Bush plan" and offered the president permission to claim it as his own. Forty Democratic senators sent their own letter the next week.

Following these intercessions, Bush renewed his promise that the United States would "play a leadership role in global warming." Even Sununu seemed to have softened. On October 30, the day before Reilly and Bromley were to leave for the Noordwijk meeting, they accompanied Sununu to the Mayflower Hotel, where he was to address international investors in the American Stock Exchange. He devoted most of the speech to explaining why there had to be a coordinated international response to the threat of climate change. When an investor asked who would pick up the cost, Sununu, with undisguised imperiousness, produced the complementary rejoinder: "Who picks up the cost if we don't fulfill our responsibilities as stewards of the environment?" Yes, he acknowledged, climate policy would incur some expense in the short term, "but if it is done constructively, the long-term cost will be less by doing it right now than it will be by trying to retreat from a disaster fifty or a hundred years from now."

But upon arrival in Noordwijk, Bromley appeared to be shrugging all this off. The Dutch were especially concerned about this development, as even a minor rise in sea level would swamp much of their nation.

Pomerance and his crew planned to stage a stunt each day to embarrass Bromley. The first took place at the flagpoles. Performing for the photographer from Agence France-Presse, Boyle and Becker lowered the Japanese, Soviet, and American flags to half-staff. Becker gave a reporter an outraged statement, accusing the three nations of conspiring to block the one action necessary to save the planet. The article appeared on front pages across Europe.

On the second day, Pomerance and Becker met an official from Kiribati, an island nation of thirty-three atolls in the middle of the Pacific Ocean about halfway between Hawaii and Australia. They asked if he was Kiribati's environmental minister.

"Kiribati is a very small place," the man said. "I'm the environmental minister. I'm the science minister. I'm everything. If the sea rises, my entire nation will be underwater."

Pomerance and Becker exchanged a look. "If we set up

a news conference," asked Pomerance, "will you tell them what you just told us?"

Within minutes they had rounded up a couple dozen journalists.

"There is no place on Kiribati taller than my head," began the minister, who was about five feet tall. "So when we talk about one-foot sea level rise, that means the water is up to my shin."

He pointed to his shin.

"Two feet," he said, "that's my thigh."

He pointed to his thigh.

"Three feet, that's my waist."

He pointed to his waist.

"Am I making myself clear?"

Pomerance and Becker were ecstatic. The minister came over to them. "Is that what you had in mind?" he asked.

It was a good start—and necessary too. Pomerance had the plunging sensation that the momentum of the previous year was under threat. The censorship of Hansen's testimony and the inexplicably strident opposition from John Sununu were unsettling signs. So were the findings of a report Pomerance had recently commissioned at the World Resources Institute, tracking global greenhouse gas emissions. The United States was the largest contributor by far, producing nearly a quarter of the world's carbon emissions, and its contribution was growing faster than that of every other country. Bush's indecision, or inattention, had already managed to delay ratification of a treaty until 1990 at the soonest, perhaps even 1991. By then, Pomerance worried, it would be too late.

The one meeting to which Pomerance's atmospheric delegation could not gain admittance was the only one that mattered: the final negotiation. The scientists and IPCC staff members were asked to leave; only the environmental ministers—and Allan Bromley—remained. Pomerance and the other activists staked out the carpeted hallway outside the conference room, waiting and thinking. Incredible as it seemed to Pomerance, it had been a decade since he helped warn the White House of the dangers posed by fossil fuel combustion; nine years since his first desperate efforts, at a fairy-tale castle in the Gulf of Mexico, to write legislation, reshape American energy policy, and demand that the United States lead an international campaign to arrest climate change. It had been a year since he devised the first emissions target proposed at a major international conference. Now, at the dawning of the decade, senior diplomats from more than sixty nations were debating the merits of a binding global treaty. But Pomerance was powerless to participate. As he stared at the wall separating him from the ministers' muffled debate, he could only hope that all his work had been enough.

The meeting began in the morning and continued into the night, much longer than expected; most of the delegates had come to Noordwijk prepared to sign the Dutch proposal. To use the bathroom, the diplomats had to exit the conference room and negotiate the hallway, squeezing past the activists; each time the doors opened and a minister darted out, the activists leapt up, demanding an update. The ministers maintained a studied silence, but as the negotiations went past midnight, their aggravation was recorded in their stricken faces and opened collars. Some time later, the Swedish minister surfaced.

"What's happening?" shouted Becker, for the hundredth time.

"Your *government*," said the minister, "is fucking this thing up!"

When, close to dawn, the beaten delegates finally emerged, Becker and Pomerance learned what had happened. Bromley, at the bidding of John Sununu and with the acquiescence of Britain, Japan, and the Soviet Union, had forced the conference to abandon the commitment to freeze emissions. The final statement noted only that "many" nations supported stabilizing emissions—but it did not indicate which nations, or at what level, or by what deadline. And with that, a decade of excruciating, painful, exhilarating progress turned to air.

The environmentalists spent the morning giving interviews and writing press releases. "You must conclude the conference is a failure," said Becker, calling the dissenting nations "the skunks at the garden party." Greenpeace called it a "disaster." In Washington, Al Gore mocked Bush on the floor of the Senate: for all the brave talk about "the White House effect," the president was practicing the "whitewash effect." The United States had proved itself "not a leader, but a delinquent partner," said Timothy Wirth. "I am embarrassed."

Pomerance tried to be more diplomatic. "The president made a commitment to the American people to deal with global warming," he told *The Washington Post*, "and he hasn't followed it up." He didn't want to sound defeated. "There are some good building blocks here," he said, and he meant it. The Montreal Protocol, the ozone agreement, wasn't perfect at first either—it had huge loopholes and weak restrictions. Once in place, however, the restrictions could be tightened. Perhaps the same could happen with climate change. Perhaps. Pomerance was not one for pessimism. As William Reilly told reporters, dutifully defending the official position forced upon him, it was the first time that the United States had formally endorsed the concept of an emissions limit. Pomerance wanted to believe that this was progress. To do so, however, he'd have to forget everything he'd learned since opening the pages of the coal report. He had been brave enough to tell the truth about Earth's future to Congress, to three presidents, to the world. But there was a limit to what he dared to tell himself.

Before leaving the Netherlands, he joined the other activists for a final round of drinks and commiseration. He knew that he would have to return to Washington the following day and start all over again. The IPCC's next policy-group meeting would take place in Edinburgh in two months, and there was already concern that the Noordwijk failure might lower the members' expectations for a treaty. But Pomerance refused to be dejected—there was no point to it. The other activists were more visibly disappointed but they shared his resolve. Alden Meyer would testify a few days later alongside Allan Bromley at a Senate hearing called by John Kerry, a Democrat from Massachusetts, to investigate why the United States had failed to sign a strong statement at Noordwijk. "I think a simple slogan might describe the situation today,"

Meyer would say. "While Bush fiddles, the earth warms." Stewart Boyle would go back to London and finish editing a report, to be published in January, showing that major cuts in carbon emissions could be achieved cheaply—"a message," he would say, "which strengthens the case for unilateral action rather than waiting for international agreements." And Daniel Becker was going to resume a Sierra Club campaign to raise automobile fuel standards. But mainly he was anxious to rejoin his wife. They had just learned that she was pregnant with their first child.

She had traveled with Becker to the Netherlands to visit friends ahead of the conference. Their hosts took them on a day trip to the southwestern province of Zeeland, where three rivers emptied into the sea. After a flood in 1953, when the sea swallowed much of the region, killing more than two thousand people, the defiant Dutch built the Delta Works, a vast fortress of movable barriers, dams, and sluice gates. It was astonishing to behold: a masterpiece of human ingenuity and imagination. All week in Noordwijk, Becker couldn't stop talking about it. The whole Delta Works could be locked into place within ninety minutes, a concrete-and-steel force field defending the land against storm surge. It reduced the country's exposure to the sea by 435 miles. The United States coastline was about 95,000 miles long. How long was the entire terrestrial coastline? Because the whole world, Becker said, was going to need this. He said that in Zeeland he had seen the future.

Afterword: Glass-Bottomed Boats

You say I'm lost,
I disagree.
The map has changed,
And with it, me.

Gliding through the seaweed,
What strange things I see below.
Cars are waiting,
Windshields wiping,
Nowhere left to go!

The ice caps are melting,
Oh ho, ho ho!
All the world is drowning,
Ho, ho ho, ho ho!
The ice caps are melting,
The tide is rushing in,
All the world is drowning,
To wash away the sin.

—TINY TIM, "THE OTHER SIDE," 1967

I asked William Reilly whether he believed that John Sununu was the only person standing in the way of a binding international agreement to prevent catastrophic global warming.

"Yes," he said. "And no."

Sununu's obstruction was critical, Reilly granted, at a time when there was little concerted opposition from any quarter, public support for climate policy was at an all-time high, the IPCC process received vocal bipartisan support, and a binding treaty along the terms broadly agreed on at Noordwijk would have kept planetary warming to 1.5 degrees.

But the negotiation of the first IPCC accord continued

for another two and a half years before it was finalized at the 1992 Rio Earth Summit, the largest gathering of world leaders in history. (Reilly led the U.S. delegation; George H. W. Bush, after some equivocation, attended.) At any point Bush could have demanded a binding treaty, and likely compelled one: the United States, after the dissolution of the Soviet Union, not only dominated the world order economically and militarily, but was responsible annually for more than one-third of humanity's carbon emissions. Bush's chief scientist, D. Allan Bromley, after his week with Reilly in Noordwijk, grew increasingly supportive of major climate policy, pushing Bush to reconsider carbon taxes and cap-and-trade plans; others in the administration joked that Reilly had brainwashed him. And in the final, critical six months of negotiation, John Sununu was powerless to protest: he resigned in December 1991, after an impressive run of effrontery that began when he was caught taking military aircraft to a dental appointment in New Hampshire.

By then, however, Bush's entire economic council had turned, having consolidated behind the position that the benefits of emissions cuts should be weighed against immediate economic costs. And those costs needed to be calculated—slowly and deliberately. At one meeting in the White House, a member of the economic council cautioned Bush against allowing environmentalists, who used a very different strain of cost-benefit analysis, to dictate the terms of a treaty. "Mr. President," he warned, "this is a bet-your-economy decision."

Even if one grants that Sununu can be blamed entirely for Noordwijk and, by extension, the thwarting of a binding global treaty, the easy success of his obstruction poses a more unsettling question: Why was support for a climate remedy so shallow that all it took was a single naysayer within the Bush administration to unravel it?

I asked Sununu the same question I asked Reilly. Had it not been for you, I asked, would we have had an effective global warming treaty? If you had argued with the same force in *favor* of a binding climate treaty, might we have one today?

"It couldn't have happened," Sununu told me, "because the leaders in the world at that time were all looking to seem like they were supporting the policy without having to make hard commitments that would cost their nations serious resources. That was the dirty little secret at the time." The IPCC process, he believes, was a face-saving act of meaningless symbolism that could lead to nothing but false promises. Even if the United States had signed a strict treaty, Sununu is convinced that it wouldn't have had any bearing on emissions levels.

Allan Bromley had evidence of this firsthand, or so he claimed in a memoir published shortly before his death in 2005. At Noordwijk, he found the lack of both economic and technical understanding "astonishing." When he asked delegates from the major European nations how they intended to stabilize their greenhouse gas emissions, they could not answer. "Who knows?" one of them told him. "After all, it's only a piece of paper and they don't put you in jail if you don't actually do it."

Thus the syllogism of binding global treaties: There is no global police force, and no appetite for economic sanctions or military action triggered by a failure to meet emissions targets, so we can only enforce ourselves. And if we are willing to enforce ourselves, what need for a binding treaty? John Sununu: "The other nations were saying we'll ride the horse, and since we don't have to make any commitments, we can look like we're on board. And frankly, that's about where we are today."

Where we are today: More carbon has been released into the atmosphere since November 7, 1989, the final day of the Noordwijk conference, than in the entire history of civilization preceding it. Earth is now as warm as it was before the last ice age, 115,000 years ago, when the seas were more than twenty feet higher. In 1990, humankind emitted more than 20 billion metric tons of carbon dioxide. In 2018, we were projected to have produced 37.1 billion metric tons—a record. Since the turn of the twenty-first century, the world's fastest-growing energy source has been coal. Despite every action taken since the Charney report—the treasure invested in research and renewable energy, the nonbinding treaties, commitments, and pledges—the only number that counts, the total quantity of emitted greenhouse gases, has continued its inexorable rise.

Our understanding of the problem hasn't changed substantially during this time. Ken Caldeira, at the Carnegie Institution for Science in Stanford, has a habit of asking new graduate students to name the largest fundamental breakthrough in climate physics since 1979. It's a trick question. There has been no breakthrough. As with any mature scientific discipline, there is only refinement. The computer models grow more precise; the regional analyses sharpen; the estimates solidify into observational data. The inaccuracies have inclined toward understatement. Caldeira's own research has shown that the world is warming more quickly than most climate models have predicted. He found that the toughest emissions reductions so far proposed will probably fail to achieve "any given global temperature stabilization target."

The political story hasn't changed greatly either, except in its particulars. Sununu is correct that, to this day, even some of the nations that have advocated most aggressively for climate policy—among them the Netherlands, Canada, Denmark, and Australia—have failed to honor their own commitments. William Nordhaus has diagnosed the problem succinctly: "Countries have strong incentives to proclaim lofty and ambitious goals—and then to ignore these goals and go about business as usual." Only seven countries are close to limiting emissions at the level necessary to keep warming to 2 degrees: India, the Philippines, Gambia, Morocco, Ethiopia, Costa Rica, and Bhutan. When it comes to the United States, which has not deigned to make any binding commitments whatsoever, the dominant narrative for the last quarter century has concerned the unrestrained efforts of the fossil fuel industry, compounded by the ingratiating abetment of the Republican Party, to suppress scientific fact, confuse the public, and bribe politicians.

The mustache-twirling depravity of these campaigns has left the impression that the industry always operated thus. But while the Exxon scientists and American Petroleum Institute clerics of the seventies and eighties were hardly good Samaritans, they did not initiate multimillion-dollar disinformation campaigns, pay scientists to prevaricate, or try to brainwash elementary school children, as their successors would. The germ of this onslaught can be traced to Jim Hansen's testimony before Congress on June 23, 1988. After Exxon's Duane LeVine consulted on strategy with a public relations officer, he gave a presentation on the greenhouse effect to Exxon's board of directors in February 1989, emphasizing "the uncertainty in scientific conclusions." LeVine predicted that global warming policy would closely follow the trajectory of atmospheric ozone policy, speculating—correctly, it would turn out—that a global treaty would be finalized in 1992. Such a treaty, he maintained, should avoid adopting "Draconian" policies that might lead to "premature limitations on fossil fuels," since the projections concerning the magnitude and timing of global warming could not be trusted. He did endorse, however, pursuing measures that reduced carbon emissions while offering other economic and environmental benefits—energy conservation, reforestation, and the development of renewable forms of energy—much as Irving Mintzer had proposed at the American Petroleum Institute the previous year, and Rafe Pomerance had proposed at the Pink Palace in 1980.

Terry Yosie drafted API's own initial "Position on Global Climate Change" in July 1989. It echoed LeVine's arguments almost exactly—warning against "premature" and potentially "counterproductive" policies "based on today's limited understanding of the issue," while advocating for "measures that will reduce the threat of climate change, yet also make sense in their own right."

In February 1990, two months after Noordwijk, these twin statements became the default position of the global petroleum industry. Exxon's LeVine, it so happened, served as chairman of the International Petroleum Industry Environmental Conservation Association's working group on climate change, the industry's liaison to the IPCC. Yosie was another member of the group; they were joined by Exxon's climate expert, Brian Flannery, and men from Shell, Texaco, BP, and Saudi Aramco. The Exxon and API position papers were included in the briefing booklet the group prepared for members of the IPCC. The industry's stance at the time was, as Yosie put it to me, echoing Henry Shaw, "defensive"; the idea was to match skepticism with accommodation, in the hope of moderating policy changes that seemed inevitable.

But a more combative approach was simultaneously being piloted, without oversight or even much forethought. Of the approximately $100,000 API budgeted to carbon dioxide policy after the Hansen hearing, a small fraction went toward establishing a lobbying organization called, in an admirable flourish of newspeak, the Global Climate Coalition. Initially run out of the offices of the National Association of Manufacturers, it was joined by the U.S. Chamber of Commerce and thirteen other trade associations, including those representing the coal, electric-grid, and automobile industries—though membership soon increased several-fold. The GCC was conceived as a reactive body, to share news of any proposed regulations, but on a whim it added a press campaign, to be coordinated by API's own communications shop, led by Charles Sandler, a veteran lobbyist, and Arthur Wiese, formerly the *Houston Post* Washington bureau chief and president of the National Press Club. They gave a few briefings to pal politicians and contacted scientists who had professed doubts about global warming. The latter included Fred

Singer and Patrick Michaels from UVA, leading skeptics during the ozone-depletion debate; and an MIT meteorologist, Richard Lindzen, the son of refugees from Nazi Germany, who shared John Sununu's anxieties about the exploitation of science by authoritarian ideologies. For an original op-ed, API offered $2,000.

In October 1989, comments from GCC scientists began to surface in national publications. They gave an issue that lacked controversy a convenient fulcrum. "You know very well we can't predict the weather with any certainty," Lindzen said "with a chuckle," in a major AP report claiming that "many respected scientists say the available evidence doesn't warrant the doomsday warnings." It was a useful refrain: the skeptics' scientific arguments were tidily debunked, but their broad characterization of the views of "the scientific community" went unquestioned. The *Times* published a letter from Singer claiming that there was "considerable doubt in the scientific community about the greenhouse climate warming." *Forbes* devoted its cover to "The Global Warming Panic," and even *Newsweek*, which had published frequent reports on the subject for two years, was prompted to ask, "Is It All Just Hot Air?" The journalistic fetish for balanced coverage was easily exploited. Despite the sudden preponderance of articles about the "doubt in the scientific community," *Science* in 1991 placed the total number of "outspoken greenhouse dissidents" in the United States at "a half-dozen or so."

Cheap and effective, GCC-like front groups began to proliferate, their cynicism laid bare by their parodic names: Citizens for the Environment, the Information Council on the Environment (ICE), the Advancement of Sound Science

Coalition, the Cooler Heads Coalition, the Global Climate Information Project, and the George C. Marshall Institute, named after the great architect of American-led multilateralism. The last, founded in 1984 to support Reagan's hawkish nuclear policy, was directed for a period by Robert Jastrow, Jim Hansen's first boss at NASA, and counted William Nierenberg, chairman of the *Changing Climate* report, among its leading members. (Nierenberg told an interviewer in 1996 that he believed serious consequences of climate change would not be felt for 150 years and expected that technological innovation would solve the problem before that, most likely through the substitution of nuclear energy for fossil fuels.) The corporations that funded these groups together represented the lion's share of the gross domestic product.

Investments in persuasion peddling rose to the level of a line item during the run-up to the 1992 Rio Earth Summit, where Bush refused for the final time to commit to specific emissions reductions. The following year, after President Bill Clinton proposed an energy tax in the hope of meeting the goals of the Rio treaty, API's number two, William O'Keefe, assumed control of the GCC and directed a $1.8 million API investment in a GCC disinformation campaign. Senate Democrats from oil and coal states joined Republicans to defeat Clinton's tax proposal, which was widely blamed for the Democrats' rout in the midterm congressional elections in 1994—the first time the Republicans had won control of both houses in forty years. Through the rest of the decade, the GCC spent at least $1 million a year to crush public support for climate policy.

The IPCC process continued, however, ahead of a 1997 summit in Kyoto. Timothy Wirth, who retired from the Senate after serving one term, led the American delegation as undersecretary of state for global affairs; he was joined by Rafe Pomerance, who had been named deputy assistant secretary in the State Department's Bureau of Oceans and International Environmental and Scientific Affairs. Every effort by Wirth's delegation to win support for emissions reductions and carbon trading brought poisonous attacks from industry and the Republican Party, coordinated by the GCC. Clinton and Gore, though supportive of Wirth's efforts, failed to win over the opposition even within their own administration, particularly among their economic advisers. Though the U.S. delegation endorsed the Kyoto Protocol—committing its parties to reduce greenhouse gas emissions in about two decades by an average of 5 percent—it was never submitted to Congress for ratification. After the GCC spent $13 million on a single ad campaign, the Senate voted on a preemptive resolution declaring its opposition to a treaty. It passed 95–0. There has not been another serious effort to negotiate a binding global climate agreement since. The closer we have come to taking action, the sharper the backlash, the more abased the retreat.

The GCC disbanded in 2002 after the defection of several major members who had grown embarrassed by its tactics; one senior Shell employee said, "We didn't want to fall into the same trap as the tobacco companies who have become trapped in all their lies." Besides, those lies weren't required anymore. George W. Bush and Dick Cheney, who for the previous five years had been CEO of the oil-services behemoth

Halliburton, had beaten Gore and won the White House. By that point, emboldened by their undefeated record against climate policy over the previous decade, the GCC and its sister groups had made a subtle, if audacious, correction to their public relations strategy. No longer did they emphasize uncertainty in the "magnitude and timing" of climate change, while ceding the potential for worst-case scenarios. Now they pushed a wilder claim: that *the fundamental science of climate change*, established by Tyndall and Arrhenius in the nineteenth century, ratified by Jule Charney's group in 1979, and confirmed by every major study since, *was itself uncertain*—a rhetorical feint akin to a historian who turns from arguing that slavery was not the primary cause of the Civil War to arguing that slavery did not exist. Though George W. Bush himself acknowledged that climate change was real and vowed (however disingenuously) to reduce greenhouse gas emissions, and during the 2008 presidential campaign the Republican nominee John McCain called for a mandatory limit on U.S. emissions, the cult of denialism overtook much of the Republican Party after Barack Obama's inauguration. The Senate, with a fifty-nine-seat Democratic majority, declined to take up comprehensive climate legislation in 2009, despite passage in the House. In that year alone, the oil and gas industry spent about half a billion dollars on lobbying efforts to weaken energy legislation.

The largest donor to that lobbying campaign was ExxonMobil, which in 2008, under pressure from shareholders—including members of the Rockefeller family, who later brought a lawsuit—had announced that it would no longer fund "public policy research groups" that advanced climate skepticism.

But ExxonMobil has continued to do so to this day, even as it places national television advertisements featuring attractive young scientists experimenting with green algae. This has made the company an especially vulnerable target for the wave of compensatory litigation that began in earnest in 2015 and may last a generation. In recent years, tort lawsuits have become a possible remedy, as the science of attributing regional effects to global emissions has grown more precise; other cases have made use of the Clean Air Act, the National Environmental Policy Act, the Endangered Species Act, the Fifth and Ninth Amendments, the Take Care Clause, the Separations of Power Doctrine, and the Public Trust Doctrine. This is one subfield of climate science that has advanced significantly since 1979—the assignment of blame.

The rallying cry of this multipronged legal effort is "Exxon knew." It is incontrovertibly true that senior employees at Exxon, and its predecessor, Humble Oil, like those at many other major oil and gas corporations, knew about the dangers of climate change at least as early as the 1950s and did nothing to reduce emissions. But it is also true that the automobile industry, responsible for nearly one-fifth of U.S. carbon emissions, knew. That industry has studied the issue since the 1970s; GM, to take one example, hosted a scientific workshop on the greenhouse effect in 1981, sent a vice president to attend the *Changing Climate* gala, and employed for most of the decade a respected atmospheric scientist, Ruth Reck, to oversee its climate change research. The electric utilities, responsible for 29 percent of national emissions, have also known since the 1970s, when their trade research association, the Electric Power Research

Institute, began conducting studies on the subject. They all own responsibility for our current paralysis and have made it much more painful than necessary. But they haven't done it alone.

The U.S. government knew. Roger Revelle began serving as a Kennedy administration adviser in 1961, five years after establishing the Mauna Loa carbon dioxide program, and every president since has debated the merits of acting on climate policy. Congress has been holding hearings for forty years; the intelligence community has been tracking the crisis even longer. The preeminent scientific journals, *Nature* and *Science*, have been publishing climate change studies for nearly a half century.

The environmentalists knew too—items appeared in the newsletters for the Sierra Club and the National Resources Defense Council in the late 1970s. With the exceptions of Friends of the Earth and the World Resources Institute, however, there was no sustained effort by activists to address the crisis until the late 1980s.

Everybody knew. In 1953, four years before Revelle and Seuss's landmark paper on humanity's "large-scale geophysical experiment," *Time*, *The New York Times*, and *Popular Mechanics* ran articles about the Canadian physicist Gilbert Plass, who had found that fossil fuels might have already warmed Earth by 1 degree Celsius. Worse was yet to come, Plass predicted, but the *Times*' science editor, Waldemar Kaempffert, saw where things were headed. "Coal and oil are still plentiful and cheap in many parts of the world," he wrote, "and there is every reason to believe that both will be consumed by industry so long as it pays to do so."

In 1956, *Time* published a profile of Revelle ("One Big Greenhouse"), questioning whether "man's factory chimneys and auto exhausts will eventually cause salt water to flow in the streets of New York and London." The same year, *Life*, with its 5.7 million circulation, published a lengthy essay about the "long-term change in world climate" that was already raising global temperatures. In 1958, the Bell Science Hour, one of the most popular educational film series in American history, aired in prime time *The Unchained Goddess*, a film about meteorological wonders, produced by Frank Capra, a dozen years removed from *It's a Wonderful Life*. In one scene, the kindly host, bald, bespectacled Dr. Research (Frank Baxter), warns his costar, the tan, tie-loosened Writer (Richard Carlson), that "man may be unwittingly changing the world's climate" through the release of carbon dioxide from factories and automobiles.

"This is bad?" asks the Writer, speaking for us all.

As Capra runs footage of glaciers collapsing like downed skyscrapers and a crude animation of a sightseeing boat tour floating above an underwater city, Dr. Research confirms that it is very bad indeed:

> A few degrees' rise in the earth's temperature would melt the polar ice caps. And if this happens, an inland sea would fill a good portion of the Mississippi Valley. Tourists in glass-bottomed boats would be viewing the drowned towers of Miami through one hundred and fifty feet of tropical water. For, in weather, we're not only dealing with forces of a far greater variety than even the atomic physicist encounters, but with life itself.

Life itself. Capra's film was shown in American science classes for decades.

Everyone knew—and we all still know. We know that the transformations of our planet, which will come gradually and suddenly, will reconfigure the political world order. We know that if we don't sharply reduce emissions, we risk the collapse of civilization. We know that 2 degrees of warming is considerably worse than 1.5 degrees, and that the use of half-degree intervals is itself euphemistic; every gradient is worse than the last: 2.1 degrees is considerably better than 2.2 degrees, which is dramatically better than 2.3 degrees. We also know that the coming changes will be worse for our children, worse yet for their children, and even worse still for their children's children, whose lives, our actions have demonstrated, mean nothing to us.

It is not easy to accept this. The equivocations spring up like oxalis after a downpour: *The situation must not be quite so terrible as that*; *surely there's time for a sensible transition to renewable energy*; *of* course *we care about our grandchildren.* But it's unseemly to cherry-pick the scientific projections or pretend that the climate will cease warming at some fixed date, fifty or one hundred years from now. The carbon cycle is ignorant of our windows and timetables, our "foreseeable future."

We do not like to think about loss, or death; Americans, in particular, do not like to think about death. No matter how obsessively one follows the politics of climate change, it is difficult to contemplate soberly an existential threat to the species. Our queasiness even infects the language we use to describe it: the banalities of "global warming" and "climate change" perform the linguistic equivalent of rolling on sanitary gloves to palpate a hemorrhaging wound. The globe and the climate will be fine, of course. They have changed drastically before and will do so again. Human beings will not be fine. Beyond an increase of 5 degrees we face the prospect of a new dark age. It is difficult to look at this fact squarely and not flinch. But doing so has a clarifying influence. It brings into relief a dimension of the crisis that to this point has been largely absent: the moral dimension, which is to say, the heart of the matter.

We are well enough acquainted by now with the political story of climate change, the technological story, the economic story, the industry story. They have been told expertly, and exhaustively, by journalists and scholars. They are all critical to understanding how we got here. But what about the human story? How does a sentient person alive now—the world already having warmed by more than 1 degree Celsius, with another 0.5-degree warming inevitable, and emissions continuing to rise unabated—how does one live with the knowledge that the future will be far less hospitable than the present? Should we obsess over it, ignore it, find some tense middle territory? What do our failures say about our substance as a people, as a society, as a democracy? Will future generations be satisfied with the answers we offer for inaction? If we vote correctly, eat vegan, and commute by bicycle, are we excused the occasional airplane ticket, the laptop, the elevators, year-round strawberries, trash collection, refrigerators, Wi-Fi, modern health care, and every other civilized activity that we take for granted? What is the appropriate calculus? How do we begin to make sense of our own complicity, however reluctant, in this nightmare? I know that I'm complicit; my hands drip crude. Hell is murky.

In the United States of America, where a growing percentage of the public regards the scientific method as vaguely sacrilegious, if not blasphemous, spiritual leaders have been divided on the significance of climate change. But the most eloquent attempt to articulate a moral vision of the issue has come from Pope Francis, in his second encyclical, *Laudato si'*, "On Care for Our Common Home." He borrows one of his central insights from Ecumenical Patriarch Bartholomew, the "Green Patriarch," the spiritual leader of Orthodox Christians. Bartholomew has called on every living person to repent for the ecological damage we have contributed, "smaller or greater, to the disfigurement and destruction of creation." The pope quotes Bartholomew at length:

> "For human beings to degrade the integrity of the earth by causing changes in its climate, by stripping the earth of its natural forests or destroying its wetlands; for human beings to contaminate the earth's waters, its land, its air, and its life—these are sins." For "to commit a crime against the natural world is a sin against ourselves and a sin against God."
>
> At the same time, Bartholomew has drawn attention to the ethical and spiritual roots of environmental problems, which require that we look for solutions not only in technology but in a change of humanity; otherwise we would be dealing merely with symptoms.

Until now we have tried to deal merely with the symptoms. We have had about as much success in treating the cancer of global warming as might any oncologist permitted to deal merely with the symptoms.

As Al Gore and Tom Grumbly understood in 1980, the climate crisis, like most human dramas, has heroes, villains, and victims. Gore himself has occupied all three roles, particularly since the 2006 release of *An Inconvenient Truth*, a tutorial and polemic that owed some of its success, and much of the intense political backlash, to his celebrity. Pope Francis and Bartholomew have acted with heroism, as have the many obscure officials, scientists, and activists who have dedicated their lives to an unpopular cause, particularly those from the ostracized communities that will suffer most from climatic changes. But any consideration of the moral dimension of the climate crisis must begin with the villains—those who have tried to bewitch an unassuming public with uncertainty, lies, and the gratuitous fantasies of denialism. The morality of these tactics can only be described as sociopathic. The rot extends, however, beyond the most cartoonish forms of denialism—the snowball brandished on the Senate floor, the "educational" videos sent to elementary schools, the actors hired by the local utility to pledge support for a new power plant at city council meetings. The failure to acknowledge the problem is itself a form of denialism: a gaslighting by omission. The moderator of a presidential debate who neglects to pose a single question about climate change; the editor who declines to devote regular coverage to the issue because there is no immediate "peg," believing that a perpetual existential threat is not sufficiently newsy; the school board that skirts the topic because it seems too political or scientific—all make their own humble contributions to the thickening of the public ignorance.

It is not yet widely understood, though it will be, that the politician who claims that climate change is uncertain betrays humanity in the same fashion as the politician who fabricates weapons of mass destruction in order to whip up support for a profiteering war. It is not yet widely understood, though it will be, that when a government relaxes regulations on coal-fired plants or erases scientific data from a federal website, it is guilty of more than merely *bowing to corporate interests*; it commits crimes against humanity. The rejection of reason—the molten core of denialism—opens the door to the rejection of morality, for morality relies on a shared faith in reason. Actions to hasten carbon dioxide emissions are the ineluctable corollary of climate denialism. Once it becomes possible to disregard the welfare of future generations, or those now vulnerable to flooding or drought or wildfire—once it becomes possible to abandon the constraints of human empathy—any monstrosity committed in the name of self-interest is permissible.

The greatest trick of the professional denialists is not to convince the world that global warming is a hoax. The world remains unconvinced; even three-quarters of U.S. voters are unconvinced. The denialists' greatest trick is to convince us that *they* are convinced—that they believe what they say. But with few exceptions—John Sununu, for one—they don't. Observe their wry equivocations, their carefully phrased evasions spoken in the defiant tones of the paid spokesperson, the bully, the flat-earther. Statements like "I believe man has an impact on the climate, but what is not completely understood is what the impact is"; and "we don't understand what the effects" of climate change are; and carbon dioxide is "not the primary knob that changes" global temperatures—these come from sworn congressional testimony by cabinet members of the current presidential administration—are exactly as honest as claims that cutting taxes on the rich will help the poor or that cigarette smoking aids digestion. The denialist does not care about winning a war of ideas, only about avoiding the appearance of amorality. If the science is uncertain, inaction is blameless.

Against such rhetoric, rational arguments are self-defeating. They only help to shuttle the discourse out of the furnace of moral reckoning and into the arid corridors of policy debate. A human problem requires a human solution. One of our most effective weapons is mortal shame. Shame may have no influence on the handymen of industry, but an appeal to higher decency can work on the human beings who vote in elections. They are still, after all, human beings.

There will eventually emerge a vigorous, populist campaign to hold to account those who did the most to block climate policy over the last forty years. The lawsuits now being pioneered against the oil and gas industry and the federal government are an initial front in this campaign, one that may seem tentative compared with the vengeance to come. Yet even the most aggressive remedies—international tribunals, truth and reconciliation commissions, reparations, asset forfeiture, nationalization of the energy industry—will not fully efface the moral stain. No judicial penalty for human tragedy can. A full reckoning requires understanding the degree to which all of our lives, even those of the moral paragons who walk among us, rely on the extraction and combustion of long-buried organic matter—of the earth's ancient dead.

Since human beings discovered fire, our quality of life, measured by any available metric—longevity, health outcomes, wealth, educational attainment—has advanced in near lockstep with our energy consumption. Until now, most of that energy has derived from fossil fuels. To the extent that we are an intelligent species, aware of our past and possessing the luxury to contemplate the future, we are beneficiaries of coal and oil and gas. Nobody who lives on the electrical grid can be let entirely off the hook; certainly not any American. The economic literature shows that, after an extreme level of socioeconomic development, the correlation between energy consumption and economic growth finally breaks down; this has happened throughout the Western world, including in the United States, where carbon emissions have broadly stabilized. Nevertheless a homeless American today consumes twice as much energy as the average global citizen.

We all have a stake in the survival of civilization. Our individual stakes are not, however, equally allotted—not yet. The relationship between those who have burned the most fossil fuels and those who will suffer the most from a warming climate is perversely inverted. The inversion is both chronological (younger generations pay for their elders' emissions) and socioeconomic (the poor suffer what the rich deserve). This, too, has been understood since the 1970s. The greatest victims will be the world's most impoverished, particularly in those nations that have not yet enjoyed the benefits of industrial energy consumption, and particularly those who do not have white skin, who will disproportionately suffer from natural disasters, declines in arable land, food and water shortages, and migratory chaos. Climate change amplifies social inequity. It disadvantages the disadvantaged, oppresses the oppressed, discriminates against the discriminated against.

In Noordwijk, when the environmental minister of Kiribati used his anatomy to demonstrate the dangers of rising seas, he did not say—did not have to say—that his nation is one of the lowest emitters of carbon dioxide on the planet, not only in aggregate but per capita. The world's island nations began years ago to make a moral plea for action. "We cannot accept that climate change be treated as an inevitability," said James Michel, president of the Seychelles, at a 2014 meeting of the Alliance of Small Island States, ahead of that year's global IPCC summit:

> We cannot accept that any island be lost to sea level rise. We do not have the economic means to build sophisticated defenses. We do not have the latest technology to better adapt to the problem . . . nor do we have the economic might to apply sanctions on those most guilty of causing the problem. But we have something that is invaluable, something that is powerful: we are the conscience of these negotiations. We stand as the defenders of the moral rights of every citizen of our planet.

A year later the foreign minister of the Marshall Islands said that the islanders' forced abandonment of their homes and cultures "is equivalent in our minds to genocide."

The American college students leading the movement

to demand a Green New Deal—an omnibus piece of legislation not unlike those proposed by Timothy Wirth and Claudine Schneider in 1988 and Barack Obama in 2008—increasingly speak in the same register as the leaders of the sinking island nations. The hundreds of students who staged a sit-in at Nancy Pelosi's office after the Democrats regained control of the House of Representatives in 2018, demanding comprehensive climate legislation, said things like: "We are angry at the cowardice of our leaders," "We are standing for our future," "Our lives are at stake." We all live on islands; some just have longer coastlines than others.

The inverted cruelties of climate change extend even to Earth's wealthiest nations. In the longer term, though, we all become impoverished. Like the economic models that chart the depreciation of the GDP against sea level rise, beyond a certain threshold, the asymptote recoils violently to the axis. There is no escaping it once the pillars of society fall—not only the pillars of the global economy, like grain production and stable international relations, but the pillars of the human spirit. An underexamined worst-case scenario is the violence done to our belief in a shared humanity. The failure to act erodes our trust in human fellowship as it does our glaciers. After another generation or two of willful neglect, who will be able to take seriously the fundamental ideals—egalitarianism, fraternity, liberty—claimed as the basis for democracy?

Our collective failure to respond to the crises heightened by rising temperatures, Pope Francis writes, "points to the loss of that sense of responsibility for our fellow men and women upon which all civil society is founded." There can be no civil society without a stable climate. There can be no stable climate without a civil society. The fight to preserve one is the same as the fight to preserve the other. If a clod be washed away by the sea, all are diminished. There can be no future unless it is understood—if not by all, then at least by a safe majority of American voters—that our future will be commonly shared.

Nearly every conversation that we have in 2019 about climate change was being held in 1979. That includes not only the predictions about degrees of warming, sea level rise, and geopolitical strife but also the speculations about geo-engineering technology, the appeals to help developing nations overcome starvation and disease without relying, as we did, on massive increases in coal consumption, and the cost-benefit analyses that always seem to favor inaction. Forty years ago, the political scientists, economists, social theorists, and philosophers who studied the slow-moving threat of climate change generally agreed that we could not be counted on to save ourselves. Their theories shared a common principle: that human beings, whether in international bodies, democracies, industries, political parties, or as individuals, are incapable of sacrificing present convenience to forestall a penalty imposed on future generations. If human beings really were able to take the long view—to consider seriously the course of human history decades or centuries after our deaths—we would be forced to grapple with the transience of all we know and love in the great sweep of time. So we have trained ourselves, whether culturally or evolutionarily, to obsess over the present, fret about the medium term, and cast the long term out of our minds, as we might spit out a poison. Adaptation to climate change, the philosopher Klaus Meyer-Abich observed in 1980, "seems to be the most rational political option." It is the option that we have pursued, consciously or not, ever since.

A major difference, four decades later, is that a solution is in hand; many solutions, in fact. They tend to involve some combination of carbon taxes, renewable energy investment, expansion of nuclear energy, reforestation, improved agricultural techniques, and, more speculatively, machines capable of sucking carbon out of the atmosphere. "From a technology and economics standpoint," Jim Hansen told me, "it is still readily possible to stay under two degrees Celsius." He has developed his own proposal, which runs a decade, arrests climate change, and saves trillions of dollars. William Nordhaus, upon winning the Nobel Prize in 2018, made the same point: "The problem is political, rather than one of economics or feasibility." We can trust the technology and the economics. It's harder to trust human behavior. "From the first time I got involved with the issue until now," Al Gore told me, "the central problem has always been that the maximum considered politically feasible still falls short of the minimum required to be efficacious. Confronted with that gap, you have two options. One is to curl up into a fetal position and fall into despair. The other is to develop a strategy for expanding the limits of what is politically feasible." The gap remains, but Gore thinks it is shrinking—he credits "dramatic changes in technological innovation and business philosophy"—and he believes, despite irreparable damage having already been done, that "we now really do have a chance to overtake the problem." Nordhaus and Hansen are less optimistic. They doubt that we will keep warming below 2 degrees.

When Rafe Pomerance feels despondent, he wears a bracelet that his granddaughter made for him, to remind him why he continues to fight. He has devised his own practical solution to climate change—not a technological solution but a political one. He argues that the critical legislative body for curtailing global emissions is the U.S. Congress. If it insists on major climate policy, he believes, the rest of the world will follow. How, then, to motivate congressional action? It is the problem he has been working on, more or less, since he met Gordon MacDonald in 1979. Pomerance is now a consultant for ReThink Energy Florida, which hopes to alert the state to the threat of rising seas. Republican congressmen in Florida have a healthier fear of climate change than their colleagues—a reasonable position in the state that, by a wide margin, is most imperiled by sea level rise. Pomerance believes that if he can persuade Florida Republicans to demand policy action, they can help turn the rest of their party.

If the prospect of a wholesale political conversion seems delusional, consider that we have solved, or at least endeavored far more seriously to solve, major social crises before, some of them existential in nature. When popular movements have managed to transform public opinion in a brief amount of time, forcing the passage of major legislation, they have done so on the strength of a moral claim that persuades enough voters to see the issue in human, rather than political, terms. We do not hesitate to summon moral arguments in debates about racial injustice, nuclear proliferation, gun violence, immigration, same-sex marriage, or the accelerating rate of mechanization. Yet the public discussion of climate change rarely ventures beyond political, economic, and legal considerations. If we speak about climate as only a political issue, it will suffer the fate of all political issues. If we speak about climate as only an economic issue, it will suffer the fate of all moral crises subjected to cost-benefit analysis. The first requirement is to speak about the problem honestly: as a struggle for survival. This is the antithesis of the denialist approach. Once the stakes are precisely defined, the moral imperative is inescapable.

The cost-benefit analysis is rapidly shifting; the distant perils of climate change are no longer very distant. Many now occur regularly, flagrantly. Each superstorm and superfire is a premonition of more terrifying convulsions to come. But disasters alone will not revolutionize public opinion in the remaining time allotted to us. It is not enough to appeal to narrow self-interest; narrow self-interest, after all, is how we got here. Tens of millions of Americans who have no reason to believe that flames will lick at their patio doors or that floodwaters will surge up their driveways must still be moved to demand a full transformation of our energy system, our economy, ourselves.

The alternative is to wait for the suffering to become unbearable. Should we pursue the status quo for the next dozen years—the amount of time that the IPCC gives us to limit warming to 1.5 degrees—the fears of young people will continue to grow, in pace with the multiplying tragedies of a warming world. At some point, perhaps not very long from now, the fears of the young will overwhelm the fears of the old. Sometime later, the young will amass enough power to act. If we wait that long, there may be time yet to avoid the most apocalyptic scenarios, but little else.

Everything is changing about the natural world and everything must change about the way we conduct our lives. It is easy to complain that the problem is too vast, and each of us is too small. But there is one thing that each of us can do ourselves, in our own homes, at our own pace—something easier than taking out the recycling or turning down the thermostat, and something more valuable. We can call the threats to our future what they are. We can call the villains villains, the heroes heroes, the victims victims, and ourselves complicit. We can realize that all this talk about the fate of Earth has nothing to do with the planet's tolerance for higher temperatures and everything to do with our species' tolerance for self-delusion. And we can understand that when we speak about things like fuel-efficiency standards or gasoline taxes or methane flaring, we are speaking about nothing less than all we love and all we are.

A Note on the Sources

This history relied on the generosity of the following people, some of whom sat for more than a dozen hours of interviews, as well as additional phone calls and correspondence that lasted more than two years: Rafe and Lenore Pomerance, James and Anniek Hansen, Jesse Ausubel, William Reilly, John Sununu, Terry Yosie, Gus Speth, Al Gore, Timothy Wirth, David Durenberger, Wallace Broecker, David Harwood, George Woodwell, William Clark, Wendy Torrance, James Bruce, Daniel Becker, Irving Mintzer, D. James Baker, Akio Arakawa, Tom Grumbly, Robert Chen, Taro Takahashi, Thomas Jorling, John Topping, Curtis Moore, Michael Boskin, Edward Strohbehn, Jr., David Hawkins, Ken Caldeira, Michael MacCracken, Jimmie Powell, Betsy Agle, John Perry, Ronald Rudolph, Anthony and Helen Scoville, Peter Schwartz, William O'Keefe, E. Bruce Harrison, John Williams, James Baker III, Andy Lacis, Philip Shabecoff, Michael Glantz, Eugene Bierly, Carl Wunsch, Elbert Friday, Nick Conger, Stacie Paxton Cobos, Jonathan Jarvis, Dan Klotz, Jimmie J.

Nelson, Roger Dower, Nicky Sundt, Karl Braithwaite, David Rind, Richard Morgenstern, Anthony Del Genio, Lonnie Thompson, Allan Ashworth, Keith Mountain, Jon Riedel, Henry Brecher, David Elliott, Lisa East, Martin Hoerling, Robert Krimmel, Michael McPhaden, Tad Pfeffer, Daniel Fagre, Shad O'Neel, Richard Meserve, Eugene Skolnikoff, Lawrence Linden, Alan Miller, Benjamin Cooley, William Sprigg, Sylvia Laurmann, and Kathy Schwendenman; Ann Finkbeiner, author of *The Jasons: The Secret History of Science's Postwar Elite*; James Rodger Fleming, author of *Historical Perspectives on Climate Change*; Janice Goldblum at the National Academy of Sciences; Justin Mankin, professor of geography at Dartmouth College; Julia Olson of Our Children's Trust; Laura Kissel at Ohio State's Byrd Polar and Climate Research Center; Kevin Krajick at the Lamont-Doherty Earth Observatory; and Amanda Kistler at the Center for International Environmental Law. Additional support was provided by the staffs of the National Archives and Records Administration, the Jimmy Carter Presidential Library and Museum, the Ronald Reagan Presidential Library and Museum, the Geisel Library Special Collections at UC San Diego, UCLA Library Special Collections, the W. R. Poage Legislative Library at Baylor University, and the Brookhaven National Laboratory's Media Communication Division.

I drew from the anthropologist Myanna Lahsen's outstanding research on the history of climate denialism, as well as her use of the term *Mirror Worlds* to describe general circulation models (following Paul N. Edwards and, in a slightly different context, David Gelernter); the investigative and research work of *Inside Climate News*, the *Los Angeles Times*, Climate Files, the Center for International Environmental Law, and Benjamin Franta; and the impressive collection of oral history interviews conducted by the American Institute of Physics. The following books deeply informed my discussion of industry's involvement in climate politics: *Censoring Science* by Mark Bowen; *The Discovery of Global Warming* by Spencer R. Weart; *Merchants of Doubt* by Naomi Oreskes and Erik M. Conway; *The Heat Is On* by Ross Gelbspan; *Climate Cover-Up* by James Hoggan with Richard Littlemore; and Stephen Schneider's *Science as a Contact Sport*.

Acknowledgments

This work would not have been possible without the generous support of the Pulitzer Center on Crisis Reporting and its director, Jon Sawyer.

I am grateful for the guidance, wisdom, and unstinting encouragement of a quadrumvirate of brilliant editors: Sean McDonald at MCD, and, at *The New York Times Magazine*, which published an earlier version of this history, Claire Gutierrez, Bill Wasik, and Jake Silverstein.

Rigorous, patient fact-checking, copyediting, and proofreading was conducted by Nandi Rodrigo, Steven Stern, Lia Miller, Bill Vourvoulias, Riley Blanton, David Ferguson, Christian Smith, Susan Gonzalez, Stephanie Butzer, Maureen Klier, Rebecca Caine, Judy Kiviat, and Janet Renard. Any remaining errors are mine.

I owe a tremendous debt to the counsel and support of Elyse Cheney, Howie Sanders, Alex Jacobs, Brooke Ehrlich, Claire Gillespie, Tara Timinsky, Daniel Vazquez, and Isabel Mendia.

Everything else I owe to Meredith Angelson.